# The Plants of Lebialem Highlands, (Bechati-Fosimondi-Besali) Cameroon

## A Conservation Checklist

Yvette Harvey, Barthélemy Tchiengué and Martin Cheek

Royal Botanic Gardens, Kew
IRAD-National Herbarium of Cameroon

Kew Publishing
Royal Botanic Gardens, Kew

Kew

PLANTS PEOPLE
POSSIBILITIES

First published in 2010 by
Royal Botanic Gardens, Kew,
Richmond, Surrey, TW9 3AB, UK
www.kew.org

ISBN 978-1-84246-399-4

British Library Cataloguing in Publication Data
A catalogue record for this book is available from the British Library

Design and typesetting by Christine Beard
Publishing, Design & Photography
Royal Botanic Gardens, Kew

Front cover: Ntoo forest at Fosimondi Forest, submontane forest of *Santiria trimera*, *Cola verticillata*
and *Chassalia laikomensis*. Forest canopy, by Aline Horwath.
Back cover: Lebialem Highlands, Cameroon, decrease in forest extent, by Susana Baena.

Printed in the UK by Hobbs the Printers

For information or to purchase all Kew titles please visit
www.kewbooks.com or email publishing@kew.org

Kew's mission is to inspire and deliver science-based plant conservation worldwide, enhancing
the quality of life

# CONTENTS

List of figures . . . . . . . . . . . . . . . . . . . . . . . . . . . . . . . . . . . . . . . . . . . . . . . . . . . . 4

Preface *Louis Nkembi* . . . . . . . . . . . . . . . . . . . . . . . . . . . . . . . . . . . . . . . . . . . . . . 5

Acknowledgements *Martin Cheek & Yvette Harvey* . . . . . . . . . . . . . . . . . . . . . 7

New Names *Martin Cheek* . . . . . . . . . . . . . . . . . . . . . . . . . . . . . . . . . . . . . . . . 9

Foreword *Martin Cheek* . . . . . . . . . . . . . . . . . . . . . . . . . . . . . . . . . . . . . . . . 11

About ERuDeF *Louis Nkembe* . . . . . . . . . . . . . . . . . . . . . . . . . . . . . . . . . . . 13

The Checklist Area *Martin Cheek* . . . . . . . . . . . . . . . . . . . . . . . . . . . . . . . . . 15

Vegetation *Martin Cheek* . . . . . . . . . . . . . . . . . . . . . . . . . . . . . . . . . . . . . . . 17

Geology, Geomorphology, Soils & Climate *Martin Cheek* . . . . . . . . . . . . . . . 29

Threats to the Lebialem Highlands *Barthélemy Tchiengué* . . . . . . . . . . . . . . . 31

The Evolution of this Checklist *Martin Cheek* . . . . . . . . . . . . . . . . . . . . . . . . 33

History of Botanical Exploration of the Lebialem Highlands
*Martin Cheek, Iain Darbyshire, Laura Pearce & Xander van der Burgt* . . . . . . . . . . . . 35

Endemic, Near-Endemic & New Taxa at Lebialem Highlands *Martin Cheek* . . . . . . . . . . . . 51

*Argocoffeopsis fosimondi* A New Species *Martin Cheek & Barthélemy Tchiengué* . . . . . . . . . . 53

*Myrianthus fosi* A New Species *Martin Cheek & Jo Osborne* . . . . . . . . . . . . . . . . 59

Red Data Taxa of Lebialem Highlands *Martin Cheek* . . . . . . . . . . . . . . . . . . . 65

Figures . . . . . . . . . . . . . . . . . . . . . . . . . . . . . . . . . . . . . . . . . . . . . . . . . 90–97

Bibliography . . . . . . . . . . . . . . . . . . . . . . . . . . . . . . . . . . . . . . . . . . . . . . . 99

Read This First: Explanatory Notes to the Checklist *Yvette Harvey* . . . . . . . . . 101

Vascular Plant Checklist

    Angiospermae . . . . . . . . . . . . . . . . . . . . . . . . . . . . . . . . . . . . . . . . . . . . 107
        Dicotyledonae . . . . . . . . . . . . . . . . . . . . . . . . . . . . . . . . . . . . . . . . 107
        Monocotyledonae . . . . . . . . . . . . . . . . . . . . . . . . . . . . . . . . . . . . 150

    Gymnospermae . . . . . . . . . . . . . . . . . . . . . . . . . . . . . . . . . . . . . . . . . . 161

    Pteridophyta . . . . . . . . . . . . . . . . . . . . . . . . . . . . . . . . . . . . . . . . . . . . 162

Index . . . . . . . . . . . . . . . . . . . . . . . . . . . . . . . . . . . . . . . . . . . . . . . . . . 165

# LIST OF FIGURES

Page no.

Fig. 1  The Cameroon Highlands, showing Lebialem.  Reproduced with
permission of BirdLife International from McLeod (1987)    10

Fig. 2  The Lebialem Highlands in context showing the study area.
Based on 1: 200,000 Mame & Bafoussam sheets — aerial coverage 1963–1964    14

Fig. 3  Profile of the Lebialem Highlands study area (vertical scale exaggerated)    15

Fig. 4  Vegetation Map of the Lebialem Highlands showing surviving natural vegetation
(modified from Letouzey 1985)    16

Fig. 5  *Argocoffeopsis fosimondi* (Rubiaceae) by A. Brown    57

Fig. 6  *Myrianthus fosi* (Cecropiaceae) by A. Brown    60

Fig. 7  *Nuxia congesta* (Buddlejaceae) by S. Ross-Craig    90

Fig. 8  *Lobelia columnaris* (Campanulaceae) by W. E. Trevithick    91

Fig. 9  *Helichrysum cameroonense* (Compositae) by W. Fitch    92

Fig. 10  *Geranium arabicum* subsp. *arabicum* (Geraniaceae) by W. E. Trevithick    93

Fig. 11  *Coffea montekupensis* (Rubiaceae) by S. Dawson    94

Fig. 12  *Geophila afzelii* (B) & *G. obvallata* subsp. *obvallata* (A) (Rubiaceae)
by F. N. Hepper    95

Fig. 13  *Pentaloncha* sp. nov. of Kupe Bakossi checklist (Rubiaceae) by S. Dawson    96

Fig. 14  *Clausena anisata* (Rutaceae) by W. E. Trevithick    97

# PREFACE

The Lebialem Highlands of Cameroon, covering a surface area of 1223 km$^2$, is a region that had never had any thorough scientific surveys in spite of its ecological importance to science, conservation and development. Though the biodiversity richness of the Lebialem Highlands had been hidden over the years, its importance is gradually being recognised even though high human pressure is causing the rapid fragmentation of the landscape and destruction of the last ecological niches of endemic species. In order to help combat the loss of biodiversity and preserve the fragile ecology intact, ERuDeF began a series of studies and education programs in this area in 2001.

The Lebialem Highlands is a key conservation priority area due to its smooth transition from the lowland rainforest, piedmont forest, submontane and savannah vegetation in the area i.e. from an elevation of 200 m at Bechati to an elevation of 2500 m above sea level at Mt Magha. There is a huge diversity of flora and fauna. Situating long-term studies of forest dynamics and forest succession in this previously unknown 'biodiversity hotspot' within the region referred to as the "Biafran forests and highlands" could provide a major insight into the ecology of the Lebialem Highlands/Mt Bamboutos. This entire region within the "Biafran forests and highlands" has been jointly considered as a 'centre of endemism and biodiversity' at both continental and global levels for a wide variety of taxa including, but not limited to primates, birds and trees exist in the region (Bergl *et al.* 2006). Recent studies conducted in the region suggest that the existing protected area system (non-existent in the Lebialem Highlands area) provides a poor coverage of the montane habitats and endemic taxa and that there is need to expand the protected area network coverage to include these highland sites (Bergl *et al.* 2006). Many of these important highland sites (including the Lebialem Highlands) are under intense pressure of habitat loss and hunting and need immediate and urgent conservation and protection actions.

Very little progress had been made on the survey and taxonomy of plants, reptiles, amphibians and butterflies. However, a series of recent rapid botanical surveys (2004, 2005, and 2006) each lasting for no more than one week, have been conducted with the assistance of teams from Earthwatch Europe, Royal Botanic Gardens, Kew and the National Herbarium of Yaoundé, Cameroon. Preliminary results from these rapid surveys and on-site identification by teams of Royal Botanic Gardens, Kew have revealed an increasing number of species rare to science, red data plants, and equally, an interesting number of range-extension species crossing over from the Bamenda Highlands to the North and Bakossi/Mwanenguba Mountains to the South (report to ERuDeF by Darbyshire, 2004).

Furthermore, as this book makes clear, the significant number of Red Data species recorded in such a brief collecting period highlights the conservation importance of this site. Moreover, these collections, in the main, only consider the upper part of the submontane element of the forest; the primary lowland forest at this site is likely to produce a range of other threatened taxa interesting to science when it is more fully surveyed. In addition, the findings of this book are that the Fosimondi submontane forest contains a cross-over of taxa from both the Bamenda Highlands and Kupe-Bakossi-Mwanenguba, and is a significant site for recording range limits of rare taxa and the reasons behind these.

With only four rapid plant surveys conducted and results of additional discoveries being found with every survey, often Red Data species or species new to science, it is possible that after the completion of more comprehensive plant surveys in future, this site could become another "Centre for Plant diversity" following the Bakossi Mountains which with over 2400 plant species, 82 strict endemic species and 232 Red Data plants surpasses all other Centres of Diversity in tropical Africa (Cheek *et al.* 2004).

Louis Nkembi
President/CEO
ERuDeF
Nyenty Building Apartment 2,
Behind Job Storey Building,
Malongo St., Molyko-Buea,
SW Region, Cameroon

## REFERENCES

Bergl, R.A., Oates, J.F. & Fotso, R. (2007). Distribution And Protected Area Coverage Of Endemic Taxa In West Africa's Biafran Forests And Highlands. *Biological Conservation* 134: 195–208.

Cheek, M., Pollard, B.J, Darbyshire, I., Onana, J.-M. & Wild, C. (eds) (2004). *The Plants of Kupe, Mwanenguba and the Bakossi Mts, Cameroon: A Conservation Checklist.* Royal Botanic Gardens, Kew, UK. iv + 508 pp.

# ACKNOWLEDGEMENTS

Martin Cheek & Yvette Harvey

Herbarium, Royal Botanic Gardens, Kew, Richmond, Surrey, TW9 3AE, UK

First we wish to thank those who facilitated our fieldwork at the Lebialem Highlands Forest. The Chief of Fosimondi, Fonang Festus, and its elders and forest council are thanked for receiving delegations from our three visits there in 2004, 2005 and 2006, and for supplying guides, such as Columbus, and for showing interest in the progress of our work. We also thank the Chief and Elders of Bechati for receiving our visit there in 2006 and providing the guide Jacob Nemba Nong.

Louis Nkembi, President/CEO of ERuDeF the leading conservation NGO seeking to protect natural habitat in the Lebialem Highlands and beyond (see "About ERuDeF") is thanked, for requesting our botanical support and aiding operations in the field by sending members of his team to join our survey teams. They helped greatly by liaising with local communities and assisting field studies in the forest: Terence Atem, Denis Ndeloh and Pius Khumbah all helped in this capacity in successive years.

Jean-Michel Onana, head of IRAD-National Herbarium of Cameroon, Yaoundé is thanked for the long collaboration of his institute with RBG Kew in Cameroon, and in particular for arranging permits and formalities for fieldwork, and allowing his researchers Nicole Guedje, and especially Barthélemy Tchiengué for joining and often leading the fieldwork. Also for allowing Barthélemy Tchiengué, funded by RBG Kew, to spearhead identification of Lebialem plant specimens at Kew.

The Apiculture and Nature Conservation Organisation (ANCO) (formerly North West Beekeeping Assocation — NOWEBA) at Bamenda, particularly their chief, Paul Mzeka, aided our first field visit to Lebialem by allowing their botanists Kenneth Tahand Terence Suwinyi to join our first visit there in 2004.

The Bamenda Highlands Forest Project, then led by Michael Vabi, allowed use of their car and driver Ndamnsa Nginyu to reach Lebialem.

Earthwatch provided support in 2004 to our survey at Bali Ngemba from whence our first visit to Lebialem was made in 2004. We thank them for their many years of support for our surveys in Cameroon and especially to volunteer Adam Surgenor who joined that visit.

We would like to thank Julie Mafanny of Limbe Botanic Garden who facilitated Kenneth Enow's participation in our fieldwork.

Our fieldwork would not have been possible without generous sponsorship from the Darwin Initiative (September 2006 expedition), the Royal Botanic Garden Kew's Overseas Fieldwork Grant (April 2005 & Feb. 2006 expeditions) and our first visit, an Earthwatch supported expedition (April 2004).

The Darwin Initiative grant to Kew has been the main single source of funds to our 'Red List Plants of Cameroon' Project (http://darwin.defra.gov.uk/project/15034/), from which this volume is one of the outputs. We are extremely grateful for this assistance, without which the fieldwork, specimen identifications and species descriptions which form the basis of this book would not have been completed. A substantial part of the publication costs of this book were met by this grant. In particular we thank Ruth Palmer and Helen Beech for their administration of this grant at DEFRA, as well as George Sarkis, financial accountant at RBG Kew for assistance in managing the finances. The Edinburgh Centre for Tropical Forestry, particularly Eilidh Young, contracted by DEFRA to monitor the grant, are thanked for their constructive reviews of our project throughout its life.

At Kew we thank Laura Pearce and Xander van der Burgt, who with Aline Horwath (now Univ. Cambridge), Bate Oben (now in teaching) and Nina Rønsted (now Univ. Copenhagen) each took a turn with one of the survey teams at Lebialem Highlands and helped collect specimens and/or provided photographs on which this book is based.

The Science and Horticultural Publications Advisory Committee of RBG Kew agreed to contribute towards the publication costs. Ian Turner expertly reviewed the manuscript. Lydia White and John Harris are thanked for guidance on producing the Camera-ready copy from which this book was produced, and for arranging scanning-in of images and design work by Chris Beard.

At Kew we also thank David Mabberley, Keeper of the herbarium, and Simon Owens, former Keeper of the Herbarium, who has long supported our work in Cameroon, as has Rogier de Kok, Assistant Keeper for Regional teams, who has also championed our work. Eimear Nic Lughadha has also been consistently supportive of our efforts in Cameroon. We also thank Daniela Zappi and her support and guidance in her role as the previous Assistant Keeper under whom this project was run. Also to be thanked is Susana Baena for producing the satellite images for the back cover.

Determinations of the specimens gathered in the course of our fieldwork were made by ourselves and by numerous botanists, often world experts in their fields. These are credited as authors and co-authors under the individual family treatments. We sincerely thank them all and their institutions, for their work, towards producing the names which are used in this book.

Thanks goes to Stuart Cable of Kew. For the Mount Cameroon Project in 1997 and 1998, he originated the species database from which the checklist part of this book was produced. George Gosline subsequently developed the 'Cameroon specimen database', enabling us to print off data labels whilst still in the field, a real time-saver. He also developed the system by which data could be exported from the Access database using XML and XSLT into Microsoft Word format. For meticulous data entry over successive years on this database, Karen Sidwell, Suzanne White, Julian Stratton, Benedict Pollard, Emma Fenton and Harry de Voil are also to be thanked. Without the database this volume would have taken considerably longer to prepare.

Craig Hilton-Taylor, IUCN officer for Red Data, based at WCMC in Cambridge, is thanked for commenting on the Red Data assessments.

# NEW NAMES

Martin Cheek

Herbarium, Royal Botanic Gardens, Kew, Richmond, Surrey, TW9 3AE, UK

The following are published for the first time in this volume:

RUBIACEAE
*Argocoffeopsis fosimondi* Tchiengué & Cheek . . . . . . . . . . . . . . . . . . . . . . . . . . . . . . . p. 54

CECROPIACEAE
*Myrianthus fosi* Cheek . . . . . . . . . . . . . . . . . . . . . . . . . . . . . . . . . . . . . . . . . . . . p. 59

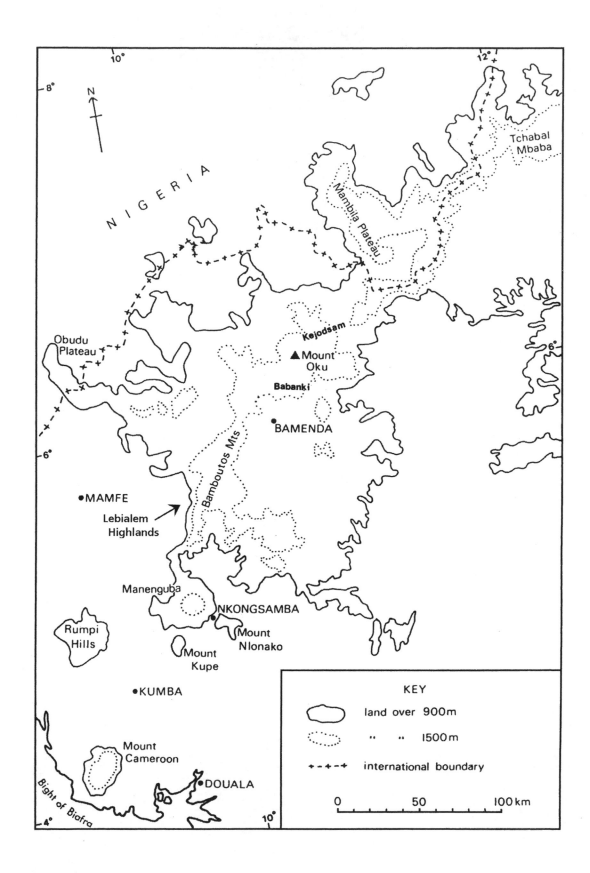

FIG. 1 The Cameroon Highlands, showing Lebialem. Reproduced with permission of BirdLife International from McLeod (1987).

10

# FOREWORD

Martin Cheek

Herbarium, Royal Botanic Gardens, Kew, Richmond, Surrey, TW9 3AE, UK

This book, while a big step forward in our knowledge of the plants of the Lebialem Highlands, is far from being the last word on the subject. It is based almost entirely on four rapid surveys by IRAD-National Herbarium of Cameroon and RBG Kew teams in 2004–2006, with the support of ERuDeF. These teams were mainly led by Dr Barthélemy Tchiengué of IRAD-National Herbarium of Cameroon. In the course of these surveys 549 specimen records were made, mainly between the villages of Fosimondi and Bechati. With the help of numerous specialists from around the world on African plant groups, spear-headed by Dr Tchiengué, 412 species have been identified from those records. Of these, six appear unique (endemic) to the Lebialem Highlands and to be new to science, occurring nowhere else in the World on current evidence. One of these, *Argocoffeopsis fosimondi* (Rubiaceae) is formally named in this volume. A further 10 probable new species to science also occur at Lebialem, of which *Myrianthus fosi* (Cecropiaceae) is also formally described here.

However, there is no doubt that many more plant species remain to be found at Lebialem. Further botanical surveys are needed to reveal the complete range of plant species present. Among the gaps remaining in our knowledge is that of the species growing between c. 700–1100 m altitude, about which we know nothing. The lowland flora is very incompletely known, and some habitats in the area, such as rock faces, have yet to be surveyed. To date only a few square kilometres have been surveyed, a very small proportion of the total area ascribed to the Lebialem Highlands by Dr Nkembi (see Preface).

The Lebialem Highlands are not just important for their numbers of endemic and threatened species. This site is also critical for understanding the distribution of species along the length of the Cameroon Highlands, filling as it does the previous large geographic gap in our knowledge between Kupe-Bakossi in the south and Bali Ngemba in the North. Surprisingly, Lebialem shares more species with the former, although geographically more distant, than with the latter, very much closer geographically.

A high proportion of the 412 species detailed in this book have been assessed as threatened with extinction, namely 42 species (see Red Data chapter), that is more than 1 in 10 of all species encountered.

It is our hope that this book will help ERuDeF to work with the Banwa people of the Lebialem Highlands to identify and protect their rarest plant species, and to avoid the threats they have faced until now, and even, with the help of forest restoration techniques, to reverse these threats.

# ABOUT ERuDeF

Very little was known about the biodiversity of the Lebialem Highlands until the advent of the Environment and Rural Development Foundation (ERuDeF) in 1999. ERuDeF is dedicated to wildlife conservation and protection of fragile environments through research, education, training and community engagement. It is the only local Cameroonian non-profit organisation that promotes the research and conservation of great apes (the Cross River gorilla and the African chimpanzee). Though focused on wildlife conservation, ERuDeF is also concerned with biodiversity conservation in general including plants and birds. Its other areas of focus are forest landscape restoration, agro-forestry, conservation education, micro-finance and enterprise development and climate change.

Over the last 10 years, ERuDeF has been able to expand its area of intervention from the Lebialem Highlands to the Lebialem-Mone Forest Landscape comprising five forest blocks namely: Bechati-Fosimondi-Besali Forest, Bechati-Mone Forest Corridor, Lebialem Highlands Montane Forest, Mak-Betchuo Forest and Nkingkwa Hills. With the support of 'Trees for the Future', ERuDeF's interventions in the domain of forest landscape restoration and agro-forestry have expanded to cover the entire western Highlands of Cameroon. While many surveys have been conducted in the wildlife section of biodiversity conservation (from 2003 till present), very little has been done in the area of plants and birds.

ERuDeF's successes over the last 10 years include but are not limited to: a complete survey of Cross River gorilla populations in the Lebialem-Mone Forest Landscape, socio-economic surveys conducted throughout the Bechati-Fosimondi-Besali forest and Bechati-Mone Forest Corridor, Conservation education implemented in many of the project area schools, preliminary survey of the plant diversity of the Bechati-Fosimondi-Besali forest by the Royal Botanical Gardens, Kew, the process of creating a Wildlife Sanctuary initiated in the Bechati-Fosimondi-Besali Forest, the project area included into the Regional Action Plans for the Conservation of Cross River gorillas and *the* Nigeria-Cameroon chimpanzees.

It is, however, hoped that through ERuDeF's long-term conservation intervention, in the near future the Lebialem-Mone Forest Landscape will encompass the following conservation units; an Important Plant Area (IPA), an Important Bird Area (IBA), a Gorilla Sanctuary, a Wildlife Corridor, a Chimpanzee Sanctuary, Mt Bamboutos Integral Ecological Reserve and a Botanical Sanctuary.

Louis Nkembi
President/CEO
ERuDeF
Nyenty Building Apartment 2,
Behind Job Storey Building,
Malongo St., Molyko-Buea,
SW Region, Cameroon

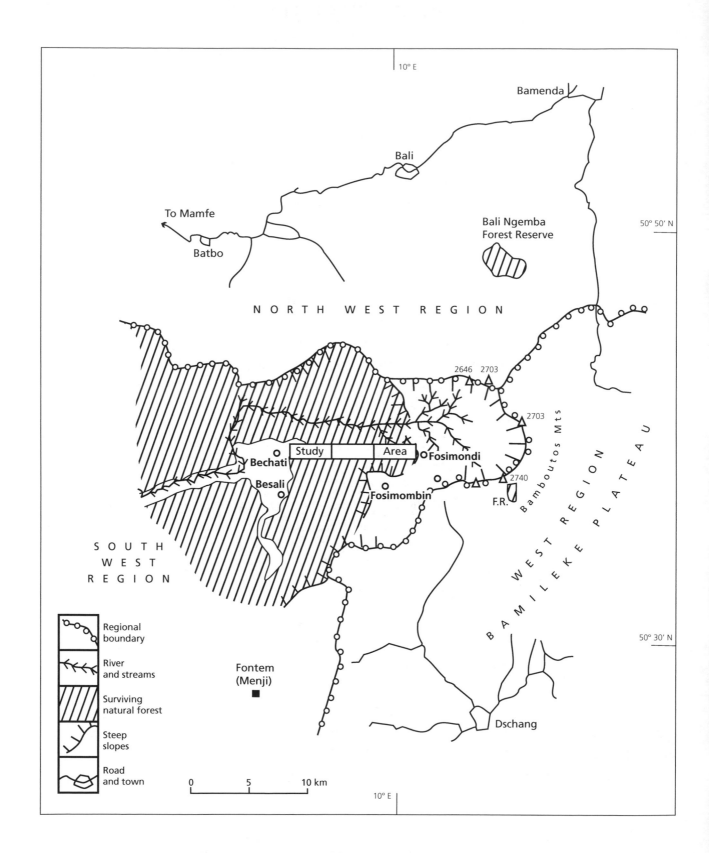

**Fig. 2** The Lebialem Highlands in context showing the study area. Based on 1: 200,000 Mame & Bafoussam sheets — aerial coverage 1963–1964.

# THE CHECKLIST AREA

Martin Cheek

Herbarium, Royal Botanic Gardens, Kew, Richmond, Surrey, TW9 3AE, UK

The checklist area consists, in effect, of a broad, and informal East-West transect (Figs. 2 & 3) through the Lebialem Highlands, situated in the extreme north-eastern corner of SW Region Cameroon. This transect has as its eastern extremity the village of Fosimondi (c. 1900 m alt., c. 5°39.5'N, 10°01'E) at the top of the steep escarpment that marks the western wall of the basaltic-trachytic Bamboutos Mts in the Cameroon Highlands line, the line being orientated roughly North–South at this juncture. At the Western extreme of our transect, at the foot of the slope, on the extremely geologically ancient and soil impoverished, basement complex, in uneven terrain, is the village of Bechati (c. 250 m alt., c. 5°40'N, 9°54'E).

While this checklist area amounts to only a few km², we believe it probably to be representative of the Lebialem Highlands as a whole, a much larger area, extending North and South of the transect, along the irregular wall of the Cameroon Highlands. The full extent of the Lebialem Highland forests is unknown, nor do they have any degree of formal protection. It is the ERuDeF project, led by Louis Nkembi, that is the main force co-ordinating the interests of local communities in protecting these forests, home to one of Africa's rarest primates, the Cross River gorilla. The triangle demarcated by the villages Besali-Bechati-Fosimondi has been the major focus of ERuDeF to date.

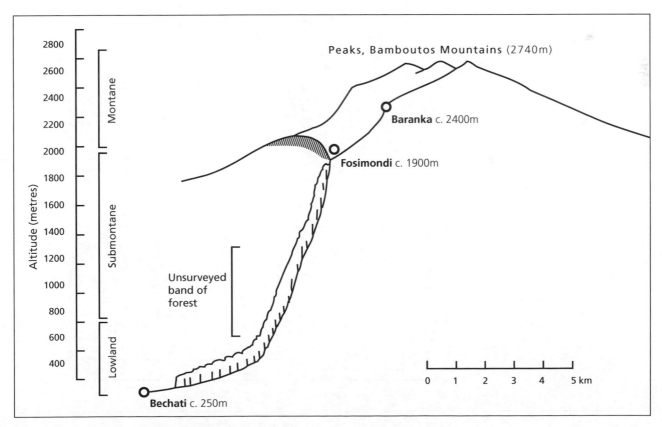

**FIG. 3** Profile of the Lebialem Highlands study area (vertical scale exaggerated).

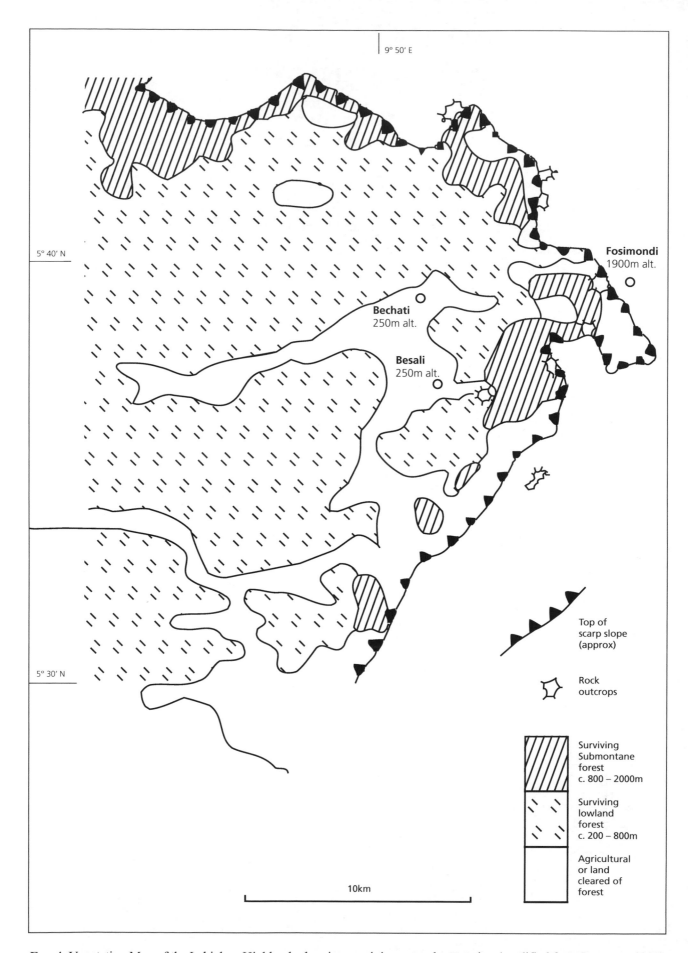

**FIG. 4** Vegetation Map of the Lebialem Highlands showing surviving natural vegetation (modified from Letouzey 1985).

# VEGETATION

Martin Cheek

Herbarium, Royal Botanic Gardens, Kew, Richmond, Surrey, TW9 3AE, UK

In this chapter the three broad natural vegetation types recognised at the Lebialem Highlands are characterised: lowland forest; submontane forest, and montane forest. Sampling by way of specimen collecting levels is gauged, the physiognomy and species composition of the vegetation is examined, endemic and threatened taxa and threats to the vegetation types are listed and phytogeographical links discussed.

## PREVIOUS STUDIES

No on-the-ground studies of vegetation in the Lebialem Highlands appear to have been conducted prior to our own. While Letouzey visited Fosimondi from Baranka in Nov. 1974, witness specimens such as *Letouzey* 13382 (*Helichrysum cameroonense*), such specimens that we have information on (*Letouzey* 13376–13393) are all of montane grassland, or scrub from altitudes higher than and to the east of our study area. This suggests that he did not survey the forests of the scarp slope that comprise the Lebialem Highlands. Letouzey (1985) produced the definitive vegetation map of Cameroon. That part pertaining to the Lebialem Highlands is the basis for our Fig. 4. Aerial photography and extrapolation from field observations made elsewhere are believed to be the basis for this excellent work.

Essentially the forests of the Lebialem Highlands in Lebialem Division of SW Region Cameroon are the steep western slopes of the Bamboutos Mountains (highest peak 2740 m) which with the Santa (2550 m), Baloum (1902 m) and Bana Mts (2097 m) dominate the Bamiléké Plateau of W Region Cameroon (itself part of the Cameroon Highland chain). According to Letouzey (1968: 337) this Plateau, average height 1000–1300 m, "covers about 5000 km$^2$, is extremely densely populated (c. 600,000 inhabitants) and is intensively cultivated (maize, peanuts, yam, [Irish] potato, macabo, taro, sweet potato, plantain etc.); its vegetation is almost entirely domestic, with some (indigenous) fruit trees (*Dacryodes edulis*, *Canarium schweinfurthii*, *Cola verticillata* etc.) and with other (native) trees and shrubs often propagated from cuttings (*Clausena anisata*, *Cola acuminata*, *Croton macrostachyus*, *Dombeya buettneri*, *Dracaena arborea*, *Ficus* spp., *Hymenodictyon floribundum*, *Markhamia lutea*, *Polyscias fulva*, *Spathodea campanulata*) supporting the wickerwork made from the rachises of *Raphia humilis* that enclose the fields" (that is forming a hedge or living fence). Fallow areas are rare (presumably the soils are fertile) and colonised by grasses. Above the plateau, at altitudes above 1900 m the slopes of the mountains are occupied by pasture; montane forest and denser scrub surviving only in the gulleys of stream valleys (Letouzey 1968: 337).

The north-east slopes of the Bamboutos Mts slope relatively gently downwards and are affected by grazing, erosion, annual burning, and in the dry season are swept by the harmattan winds from the North. In contrast the slopes to the south-west have abrupt slopes, better protected against fire and grazing and so are still forested (Letouzey 1968, citing earlier studies by Jacques-Felix). It is these forested slopes that form the main subject of this book.

In this treatment the several man-made vegetation types that occur in our area are not mapped nor characterised since they were not the subject of our study. Of the three natural vegetation types that are characterised, only the first and second, lowland and submontane forest, are mapped, since the last, montane forest is now so rare and sparse locally as to be unmappable.

## CONVENTIONS USED IN VEGETATION TREATMENTS

The species lists given have been extracted from the Checklist. Species are listed in the order in which they appear in the Checklist.

Threatened taxa (those accepted by or proposed to IUCN as 'Red Data' taxa, being either vulnerable (VU), Endangered (EN) or Critically Endangered (CR) are indicated by being annotated accordingly.

In the submontane and montane forest treatments, altitudes of records from the Checklist area are given where available since species can be confined to particular altitudinal bands within these vegetation types.

## MAPPING UNIT 1. LOWLAND EVERGREEN FOREST (c. 250–800 m alt.)

At the bottom of the precipitous western slopes of the Bamboutos Mts that form the Lebialem Highlands is found an evergreen forest belt at c. 250–800 m altitude. This area is much less densely populated than the Bamiléké Plateau above, to the East, and same areas of intact forest remain, now subject to logging. Around Bechati, briefly our base for exploring the lower slopes of the Lebialem Highlands, large areas of forest appear to have been converted long ago into *Elaeis guineensis* (oil palm) plantations which have a semi-natural appearance. Small-holder cacao (*Theobroma cacao*) and robusta coffee (*Coffea canephora*) are the main cash crops while plantains (*Musa*) and macabo (*Colocasia*) are major food crops. Rainfall exceeds 3 m p.a., and soils are mainly very poor, and sandy (see Geology chapter). Sampling of lowland evergreen forest in the Lebialem Highlands has been very low. So far as is known, the only collections that exist are those of Tchiengué and van der Burgt, made over several days in September 2006. They trekked from Bechati (c. 250 m alt.) eastwards and upslope, towards Fosimondi, following the footpath, reaching c. 650 m altitude. Much of the forest in this area had been replaced with *Elaeis*, but nonetheless they found many rare species. During their reconnaissance visit which was brief, necessitated by the very difficult access to Bechati at that season, and their limited time, they were not able to execute any plots, not to do more than make a few collections and photographs. Height of canopy and emergents was 35–50 m. No indicators of semi-deciduous forest were seen.

## Canopy Trees

*Antrocaryon klaineanum*

*Canarium schweinfurthii*

*Afrostyrax lepidophyllus*

*Napoleonaea egertonii* VU

*Hymenostegia afzelii*

*Scorodophloeus zenkeri*

*Albizia adianthifolia*

*Cylicodiscus gabunensis*

*Medusandra mpomiana*

*Sterculia tragacantha*

*Baillonella toxisperma*

*Synsepalum cerasiferum*

*Desplatsia subericarpa*

*Elaeis guineensis*

## Climbers

*Thunbergia fasciculata*

*Monanthotaxis* sp. nov. of Fosimondi

*Baissea gracillima*

*Pararistolochia* sp. of Bechati

*Marsdenia latifolia*

*Salacia lebrunii* EN

*Cnestis corniculata*

*Connarus griffonianus*

*Dichapetalum heudelotii* var. *hispidum*

*Dichapetalum tomentosum*

*Duparquetia orchidacea*

*Chlamydocarya thomsoniana*

*Polycephalium lobatum*

*Pyrenacantha longirostrata* EN

*Clerodendrum silvanum* var. *bucholzii*

*Medinilla mirabilis*

*Turraea vogelii*

*Perichasma laetificata* var. *laetificata*

*Atractogyne bracteata*

*Chassalia* sp. 1 of Bechati

*Didymosalpinx abbeokutae*

*Keetia hispida 'setosum'*

*Mussaenda tenuiflora*

*Rutidea olenotricha*

*Sherbournia zenkeri*

*Tarenna eketensis*

*Tarenna fusco-flava*

*Microcos barombiensis*

*Urera gravenreuthii*

*Cyphostemma adenopodum*

## Forest shrubs and small trees

*Rauvolfia mannii*

*Myrianthus arboreus*

*Salacia lehmbachii* var. *pes-ramulae* VU

*Jollydora duparquetiana*

*Antidesma vogelianum*

*Bridelia atroviridis*

*Mareyopsis longifolia*

*Pycnocoma cornuta*

*Leptaulus daphnoides*

*Achyrospermum oblongifolium*

*Leea guineensis*

*Leonardoxa africana* subsp. *letouzeyi*

*Angylocalyx oligophyllus*

*Heckeldora staudtii*

*Microdesmis* cf. *puberula*

*Olax gambecola*

*Olax latifolia*

*Chassalia simplex*

*Heinsia crinita*

*Lasianthus batangensis*

*Massularia acuminata*

*Oxyanthus gracilis*

*Pavetta bidentata* var. *bidentata*

*Pavetta brachycalyx* EN

*Pavetta gabonica*

*Pavetta owariensis* var. *owariensis*

*Psychotria lucens* var. *lucens*

*Rothmannia whitfieldii*

*Tarenna lasiorachis*

*Tricalysia discolor*

*Allophylus conraui* EN

*Cola ficifolia*

*Cola flaviflora*

*Cola lepidota*

*Boehmeria macrophylla*

*Rinorea dentata*

**Herbs**

*Justicia tenella*

*Stenandrium guineense*

*Celosia isertii*

*Celosia leptostachya*

*Impatiens kamerunensis* subsp. *kamerunensis*

*Begonia pseudoviola* VU

*Plectranthus epilithicus*

*Cincinnobotrys letouzeyi* EN

*Dinophora spenneroides*

*Dorstenia barteri* var. nov. of Bechati

*Argostemma africanum*

*Geophila afzelii*

*Geophila obvallata* subsp. *obvallata*

*Hymenocoleus hirsutus*

*Hymenocoleus subipecacuanha*

*Psychotria globosa* var. *globosa*

*Trichostachys petiolata* EN

*Chlorophytum comosum* var. *bipindense*

*Aneilema silvaticum* VU

*Coleotrype laurentii*

*Stanfieldiella imperforata* var. *imperforata*

*Hypolytrum heteromorphum*

*Dracaena camerooniana*

*Dracaena phrynioides*

**Rheophytic community**

*Kanahia laniflora*

*Biophytum talbotii*

*Habenaria weileriana*

The last three species listed above are restricted to fast-flowing rivers (rheophytes). Further surveying along rivers, especially along sections with rapids and waterfalls, is likely to uncover further species. Many rheophytes are of conservation importance. *Anubias barteri* var. *barteri* another rheophyte was found in streams below Fosimondi.

In conclusion, lowland evergreen forest at the foot of the Lebialem Highlands while still very incompletely known, appears to contain a high proportion of threatened species — ten are listed above. There are also three taxa probably new to science: *Monanthotaxis* sp. nov. of Bechati, *Dorstenia barteri* var. nov. of Bechati and *Chassalia* sp. 1 of Bechati.

Further botanical surveys at Bechati are likely to increase these numbers greatly as long as forest habitat remains intact in the area. On the basis of the single visit made it was not possible to gauge rate of habitat loss on the ground. Poor transport infrastructure and poor soils do not facilitate clearance of forest in the area at present.

Letouzey's vegetation map (1985) classifies the lowland forest at the foot of the Lebialem Highlands, around Bechati, as two types:

Mapping unit 203: Atlantic Biafran Forest with Caesalps (mostly below the 500 m contour).
Mapping unit 228: Atlantic forest (NE type) with Caesalps rare (above 500 m contour).

Our sampling has been so low as not to allow an attempt to assess the validity of this classification. Therefore these two types have been mapped as one.

Detailed studies of lowland forest within SW Region, Cameroon, remain uncommon, except at Korup National Park, and around Mt Cameroon. At inland sites, such as at Lebialem, the species structure of this forest type is almost completely unknown. There is an urgent need to study this habitat at such sites so as to gauge its need for protection. Generally speaking its main threat in SW Region is logging followed by conversion to agriculture.

### MAPPING UNIT 2: SUBMONTANE FOREST (c. 800–1900/2000 m alt.)

This is the key vegetation type of the Lebialem Highlands, the forest habitat that supports the Cross River gorilla, and also the largest number of threatened plant species including the endemic *Coffea* relative, *Argocoffeopsis fosimondi*. Submontane forest has been the main focus of our field surveys in the area, with surveys in April 2004, April 2005 and February 2006, all based in Fosimondi, at 1900 m alt., with teams seeking to descend on steep and slippery slopes and reach beyond the farms to altitudes where forest was still intact, sometimes using makeshift overnight camps to extend their range. While good quality forest was found at 1300–1400 m alt, the teams were not able to reach altitudes much below 1300 m, leaving a large apparently intact belt of forest, between c. 650–1270 m alt. unsurveyed apart from the visit by Jaap van der Waarde in October 2004 (see History of Botanical Exploration). Attempts to reach this band from Bechati, below, in September 2006 were also unsuccessful. Therefore the lower submontane forest of the Lebialem Highlands, which can be expected to be more species-diverse than that above, remains botanically unknown. It is to be hoped that future surveys will uncover the species that it holds.

Submontane forest is also known as cloud forest, and that in the study area is perfectly placed to intercept cloud from the prevailing south-westerly winds that bear moisture from the Atlantic Ocean since the scarp slopes on which the forest sits are arranged in an arc directed towards the south-west.

The forest canopy in cloud forest at Lebialem was generally 15–20 m above the ground, the understorey often clothed with mosses. No plots or studies of forest structure were conducted since the priority was a basic botanical survey.

While still largely intact, this vegetation type is steadily being cleared downslope from Fosimondi. Tchiengué over three annual visits, was able to observe that areas of forest studied on his first survey no longer existed on his next.

No attempt is made here to subdivide submontane forest into separate altitudinal bands as has been done at Bali Ngemba (Cheek in Harvey *et al.* 2004) to the North and at Kupe-Bakossi to the South (Cheek in Cheek *et al.* 2004). This is because:

1) the lower altitudinal belt of the submontane forest is totally unknown to science;

2) because much of the upper altitudinal band 1800–1900 m has been destroyed rendering sampling difficult;

3) where sampling has been carried out it has not been exhaustive.

## Submontane forest tree and shrub species 1290–1900 m alt.

*Alangium chinense* 1500 m

*Trichoscypha lucens* 1700 m

*Uvariodendron connivens* 1486 m

*Xylopia africana* VU

*Tabernaemontana* sp. of Bali Ngemba
1300–1600 m

*Voacanga* sp. 1 of Bali Ngemba & Kupe
1700 m

*Polyscias fulva* 1400–1800 m

*Santiria trimera* 1380 m

*Myrianthus preusii* subsp. *preusii* 590 m (error?)

*Myrianthus* sp. 1 of Kupe 1290 m

*Magnistipula conrauana* 1800 m EN

*Drypetes principum* 1350–1370 m

*Macaranga occidentalis* 1480 m

*Pseudagrostistachys africana* subsp. *africana*
1700 m VU

*Shirakiopsis* (*Sapium*) *elliptica* 1500 m

*Dasylepis racemosa* 1800 m

*Oncoba ovalis* 1350–1400 m

*Allanblackia gabonensis* 1700 m VU

*Garcinia smeathmanii* 1700 m

*Pentadesma grandifolia* 1340 m

*Symphonia globulifera* 1800 m

*Zenkerella citrina* 1800 m

*Heckeldora ledermannii* 1370 m EN

*Argocoffeopsis fosimondi* 1330–1380 m CR

*Chassalia laikomensis* 1370–1800 m CR

*Coffea montekupensis* 1290 m

*Cuviera longiflora* 1800 m

*Psychotria babatwoensis* 1355 m EN

*Psychotria camptopus* 1800 m

*Psychotria penduncularis* var. *hypsophila*
1700 m

*Psychotria succulenta*

*Psychotria* sp. 9 of Kupe-Bakossi 1300 m

*Dracaena arborea*

*Cyathea manniana* 1400 m

*Cyathea camerooniana* var. *camerooniana*
1370 m

### Epiphytic shrubs

*Schefflera hierniana* EN

*Anthocleista scandens* VU

The phytogeographic position of the submontane forest of the Lebialem Highlands is particularly interesting and important. At one time submontane forest would have been almost continuous from Mt Cameroon near the coast to Tchabal Mbabo in the North, apart from lowland areas interrupting the line of the highlands. However, submontane forest has only been sampled and studied in any detail at the species level at four sites in this line, moving northwards: Mt Cameroon (Cable & Cheek 1998), the Bakossi Mts including Mt Kupe (Cheek *et al.* 2004), Bali Ngemba (Harvey *et al.* 2004) and Dom (Cheek *et al.* 2010). It is mainly within the context of these studies and particularly the second and third, respectively c. 80 km to the S and 30 km to the N, that the following analysis is made.

## FILLING IN THE GAP

Several submontane forest previously known from Kupe-Bakossi and Bali Ngemba have now been found at Lebialem Highlands giving support to the hypothesis that they once had a continuous distribution between these two sites (and further afield in some cases) for example:

*Xylopia africana*

*Tabernaemontana* sp. of Bali Ngemba

*Begonia oxyanthera*

*Allanblackia gabonensis*

*Zenkerella citrina*

*Chassalia laikomensis*

*Cuviera longiflora*

*Tricalysia* sp. B aff. *ferorum*

*Allophylus bullatus*

*Deinbollia oreophila* (*Deinbollia* sp. 1 of Kupe & Bali Ngemba)

*Quassia sanguinea*

In contrast other several species with similar ranges have not been found yet, but this might indicate that specimen surveying at Lebialem has been insufficient. These are *Strombosia* sp. 1 and *Oncoba lophocarpa*.

## RANGE EXTENSIONS N FROM KUPE-BAKOSSI

Most exciting is the discovery in the Lebialem Highlands of several rare species previously only known from Kupe-Bakossi, 80–90 km to the S. Paradoxically this range extension has increased their threat ratings since we now know that instead of their entire population being relatively secure within the confines of Kupe-Bakossi, a large part of it extends 80 km N (or more) and must have seen large losses and continues to face threats within that northerly extension area:

*Myrianthus* sp. 1 of Kupe

*Heckeldora ledermannii*

*Coffea montekupensis*

*Pentaloncha* sp. nov. of Kupe Bakossi checklist

*Psychotria* sp. 9 of Kupe-Bakossi

*Trichostachys petiolata* (*Trichostachys* sp. 1 of Kupe Bakossi)

*Rhaptopetalum geophylax*

## RANGE EXTENSIONS S FROM BALI NGEMBA

It was unexpected that despite the proximity of Bali Ngemba (only 30 km to the N) and the high degree of altitudinal overlap with Lebialem, so few of its apparent endemics have been found to extend to Lebialem — in contrast to the case with Kupe-Bakossi. Exceptions are:

*Beilschmiedia* sp. 1 of Bali Ngemba

*Psychotria babatwoensis*

*Psychotria* sp. B of Bali Ngemba Checklist

The submontane forest of the Lebialem Highlands appears to have a much closer botanical affinity with the more distant (80–90 km) Kupe-Bakossi area than to the nearer Bali Ngemba. This suggests that there may be a natural-barrier, or partial barrier, either physical or climatic, between Lebialem and Bali Ngemba.

## THREATENED AND UNDESCRIBED TAXA

Twenty-three threatened taxa and ten that are probably new to science have been found in the submontane forest of Lebialem to-date. All of the latter ten occur also at either Kupe-Bakossi and/or Bali Ngemba except *Salacia* sp. aff. *nitida* of Fosimondi which, with *Argocoffeopsis fosimondi*, described in this volume may also be endemic to the cloud forest of Fosimondi. All ten when assessed, are likely to prove threatened.

| Threatened herbs of Submontane Forest | Undescribed species of Submontane Forest |
|---|---|
| *Impatiens letouzeyi* EN | *Tabernaemontana* sp. of Bali Ngemba |
| *Begonia oxyanthera* VU | *Voacanga* sp. 1 of Bali Ngemba & Kupe |
| *Begonia preussii* VU | *Myrianthus* sp. 1 of Kupe |
| *Begonia schaeferi* VU | *Salacia* sp. aff. *nitida* of Fosimondi |
| *Plectranthus punctatus* subsp. *lanatus* VU | *Beilschmiedia* sp. 1 of Bali Ngemba |
| *Pentaloncha* sp. nov. of Kupe Bakossi checklist 1290 m EN | *Lasianthus* sp. 1 of Kupe checklist |
| *Sabicea xanthotricha* 1400 m EN | *Mussaenda* sp. 1 of Fosimondi |
| *Bulbophyllum nigericum* VU | *Pentaloncha* sp. nov. of Kupe Bakossi checklist |
| *Cyrtorchis letouzeyi* EN | *Psychotria* sp. 9 of Kupe-Bakossi |
| *Diaphananthe bueae* EN | *Psychotria* sp. B of Bali Ngemba Checklist |
| *Polystachya bicalcarata* VU | |

## SYNCHRONOUS MASS-FLOWERING OF UNDERSTOREY MONOCARPIC HERBS IN SUBMONTANE FOREST

Simultaneous mass-flowering, followed by death, is a phenomenon recorded at Mt Oku (Cheek in Cheek *et al.* 2000: 21), and at Kupe-Bakossi (Cheek in Cheek *et al.* 2004: 60–61) and at Mt Cameroon (Cable & Cheek 1998: xvii). Of the classical four main species concerned in the Cameroon Highlands, namely *Brachystephanus giganteus* VU (*Oreacanthus mannii*), *Mimulopsis solmsii*, *Plectranthus insignis* and *Acanthopale decempedalis*, all but the last were observed in flower by Tchiengué in the April 2005–Feb. 2006 season, with the addition of *Brachystephanus jaundensis* subsp. *jaundensis* (*Tchiengué* 2610) in Feb. 2006. All of these four species were recorded in flower in the 1350–1540 m altitudinal band within this season, and none was seen in flower in the April 2004 survey. This suggests that flowering of this group of species is rather tightly restricted to a narrow window as at Mt Oku and Mt Cameroon, and is not continuous, although a few plants may be expected to flower out of synchrony.

While synchronous mass-flowering of monocarps occurs in montane forest (above 2000 m alt.) at Mt Oku (observed in 1996), it has been recorded in the *submontane* forest belt at Lebialem, as at Mt Cameroon and at Kupe-Bakossi (Cable & Cheek 1998 and Cheek *et al.* 2004). The phenomenon has not been recorded at Bali Ngemba where not a single one of the classic species has been recorded (Harvey *et al.* 2004), nor at Dom (Cheek *et al.* 2010). However, this may be because surveys at these sites did not coincide with mass-flowering years which may occur only every seven or nine years (Cheek *et al.* 2000: 21). When not in flower these species, even although locally frequent, would be easy to overlook or confuse with other Acanthaceae. At Kupe-Bakossi while mass-flowering occurred, there was little evidence of multi-species synchronicity, which is currently best recorded in the Cameroon Highlands at Mt Oku and Lebialem.

## MONTANE FOREST (1900/2000 m alt. and above)

Montane forest was once dominant and common over 2000 m alt. through the Bamenda Highlands and Bamiléké Plateau, including the Bamboutos Mts. Such forest almost certainly once extended from Fosimondi to the highest peaks of the Bamboutos Mts but now it has been almost entirely replaced by cultivation. In 1977 Letouzey documented the occurrence of the very rare montane forest tree which he named *Ternstroemia polypetala* but which we know to be a new species to science restricted to these highlands. Letouzey found it about 5 km down the road from Baranka to Fosimondi, just outside our study area. Tchiengué and his teams searched along this road for the species in 2004, 2005 and 2006 but it has not been refound and is locally, perhaps globally extinct. In concluding his scientific paper announcing the discovery of this species in Cameroon Letouzey (1977) lamented the likely fate of this and other rare mountain forest species, as follows (translated from the French):

"It is dreaded that many of these discoveries will soon be annihilated. For some years at the Mts Bamboutos, and Mt Oku and a good number of other lesser massifs, the fragments and other patches of submontane (about 1000–2000 m) and montane (c. 2000–3000 m) forests have been seriously attacked by farmers, often for subsistence agriculture, beans, cabbages, potatoes, cocoyams or for cash crops (arabica coffee and tobacco). At several points the cultivated areas have already given way to fallow areas and succession, then to pastures and the dismal and thin carpet of *Sporobolus africanus*, extending further each day, replacing the forests which were once protectors of a specific micro-climate and above all maintaining the slopes and covering the soils with all the benefits that these actions have on the water regime. This destruction, justified by very short term economic interests is perpetuated and perpetuates itself still, despite the advice and warnings given."

Today the fragments that survive are even smaller and sparser than in Letouzey's day. Tchiengué's teams visited some surviving stands of trees but found that they had mainly been burnt. Nonetheless an attenuated complement of the montane forest species seen elsewhere in the Cameroon Highlands, such as at Mt Oku (where relatively well protected: Cheek *et al.* 2000: 19–21) survive, often as isolated individuals. Many of these species were detected at altitudes below that at which they usually occur (1900–2000 m and above) on the upper part of the scarp slope below Fosimondi, where forest is more intact than in the basin of the Upper Manyu above Fosimondi.

### Montane forest canopy trees present at Fosimondi

*Carapa grandiflora* (*C. oreophila* ined.) 1800 m      *Prunus africana*

*Bersama abyssinica* 400–1400 m      *Clausena anisata* 1500 m

*Syzygium staudtii* 1400 m      *Podocarpus milanjianus* 1920 m

*Cassipourea malosana* 1700 m

Species notably unrecorded but usual in this vegetation type were: *Arundinaria alpina*, *Schefflera abyssinica*, *S. mannii*, *Rapanea melanophloeos*, *Ixora foliosa*, *Olea capensis*, *Ilex mitis* and *Ficus oreodryadum*.

## Montane forest understorey tree and shrub species recorded at Fosimondi

*Xymalos monospora* 1800 m

*Ardisia kivuensis* 1700 m

*Pittosporum viridiflorum 'mannii'* 1800 m

*Pavetta hookeriana* var. *hookeriana* 1700–
    1800 m VU

*Allophylus bullatus* 1800 m VU

*Brucea antidysenterica*

*Discopodium penninervium*

Species not recorded but expected: *Maytenus buchananii, Rytigynia neglecta, Psydrax dunlapii, Cassine aethiopica, Tarenna pavettoides, Rhamnus prinoides, Deinbollia* sp. 2 of Kupe.

## Liana species present:

*Embelia schimperi* 1300 m

Absent: *Embelia mildbraedii, Clematis hirsuta, C. simensis, Jasminum dichotomum, Stephania abyssinica.*

## Montane forest understorey epiphytic herbs

*Impatiens sakeriana* 1700 m VU

*Begonia oxyanthera* 1460–1800 m VU

*Begonia poculifera* var. *poculifera* 1800 m

## Montane forest edge trees, shrubs and herbs present:

*Croton macrostachyus* 1400 m

*Lobelia columnaris* 1700 m

*Geranium arabicum* subsp. *arabicum* 1800 m

*Psorospermum* spp.

*Nuxia congesta*

*Maesa lanceolata*

*Thalictrum rhyncocarpum* subsp.
*rhyncocarpum* 1800 m

*Rubus apetalus* 1800 m

*Galium simense* 1800 m

*Otomeria cameronica*

*Gnidia glauca*

Absent: *Neoboutonia mannii, Hypericum* spp., *Cyathea dregei, Agarista salicifolia, Morella arborea.*

## OTHER MONTANE HABITATS

Our surveys in the Lebialem Highlands have been forest-focussed, and focussed especially on the submontane forest of the scarp slope. Other habitats above Fosimondi have not been surveyed but are present although outside the study area so are not treated in detail here:

## Montane grassland:

*Helichrysum cameroonense* 2150 m (Letouzey 13382)

This rare species suggests that this habitat in this area is diverse and species rich.

## Montane scrub:

*Adenocarpus mannii*

*Clutia kamerunica*

These have been collected on the road from Baranka to Fosimondi, outside the checklist area. These indicate species-rich, high quality, montane scrub, but how much remains, and how fragmented it is, is unknown.

Sampling of montane forest and other montane habitats above the Lebialem Highlands has been fragmentary since the main focus of our study has been the submontane forest below Fosimondi. Undoubtedly, more detailed surveys of the gullies and scattered individual trees above Fosimondi, towards the highest point of the Bamboutos, would reveal more of the species of montane forest so far unrecorded. Sadly, this vegetation type has been so extensively destroyed for agricultural reasons that physiognomic characterisation is not possible. The phytogeographical links of montane vegetation above Fosimondi are with other areas above c. 2000 m alt. in the Cameroon Highlands in the first place, and secondly with other montane areas in Africa, these being East and South of the Congo basin apart from the highest points of the Guinea Highlands in West Africa, where a highly depauperate fraction of these species can be found.

Of the montane species recorded in our checklist, only four are accepted as threatened, all as VU: *Pavetta hookeriana* var. *hookeriana*, *Allophylus bullatus*, *Impatiens sakeriana* and *Begonia oxyanthera*. All four being endemic to montane forest habitat of the Cameroon Highlands. No montane species is known which is endemic to the Fosimondi area.

## REFERENCES

Cable, S. & Cheek, M. (eds) (1998). *The Plants of Mount Cameroon: A Conservation Checklist.* Royal Botanic Gardens, Kew, UK. lxxix + 198 pp.

Cheek, M., Onana, J.-M. & Pollard, B.J. (eds) (2000). *The Plants of Mount Oku and the Ijim Ridge, Cameroon: A Conservation Checklist.* Royal Botanic Gardens, Kew, UK. iv + 211 pp.

Cheek, M., Pollard, B.J, Darbyshire, I., Onana, J.-M. & Wild, C. (eds) (2004). *The Plants of Kupe, Mwanenguba & the Bakossi Mts, Cameroon: A Conservation Checklist.* Royal Botanic Gardens, Kew, UK. iv + 508 pp.

Cheek. M., Harvey, Y. & Onana, J.-M. (eds) (2010). *The Plants of Dom, Bamenda Highlands, Cameroon: A Conservation Checklist.* Royal Botanic Gardens, Kew, UK. iv + 162 pp.

Harvey, Y., Pollard, B.J., Darbyshire, I., Onana, J.-M. & Cheek, M. (eds) (2004). *The Plants of Bali Ngemba Forest Reserve, Cameroon: A Conservation Checklist.* Royal Botanic Gardens, Kew, UK. iv + 154 pp.

Letouzey, R. (1968). Les Botanistes au Cameroun. *Flore du Cameroun* 7. Museum National d'Histoire Naturelle, Paris

Letouzey, R. (1977). Presence de *Ternstroemia polypetala* Melchior (Theacees) dans les montagnes camerounaises. *Adansonia* 17 (1): 5–10

Letouzey, R. (1985). Notice de la carte phytogéographique du Cameroun au 1: 500,000. IRA, Yaoundé, Cameroon.

# GEOLOGY, GEOMORPHOLOGY, SOILS & CLIMATE

Martin Cheek

Herbarium, Royal Botanic Gardens, Kew, Richmond, Surrey, TW9 3AE, UK

The Lebialem Highlands are the western and south-western facing scarp slopes below the High Lava Plateau on which sits the Bamboutos Mts. These are the highest point (2740 m) in the Cameroon Highlands between Mt Oku (3011 m) to the North, and Mt Cameroon (4095 m) to the South (Fig. 1). The mountainous areas that comprise the Cameroon Highlands begin with the Atlantic Ocean islands of São Tomé, Príncipe, Annobón and Bioko, and proceed NE through Cameroon in a band 50–100 km wide. These mountains have their origin in a geological fault and are formed largely of igneous material. Three main periods of volcanic activity and one of plutonic uplift have been reviewed by Courade (1974). The Bamenda Highlands and Bamiléké Plateau, including the Bamboutos Mts, were largely formed in the Tertiary, in the second of the three main periods of volcanic activity, 'the middle white series'. The High Lava Plateau on which Fosimondi rests at 1900–2000 m alt. is composed of basalts and trachytes which have given rise to humid volcanic soils (Hawkins & Brunt 1965 soil map; Courade 1974). These have average fertility (Courade 1974). In contrast, at the foot of the scarp slope, where Bechati rests, at 200–300 m alt., the geology is that of ancient decomposed basement complex rocks which have given rise to sandy soils of the lowest class of fertility. Courade (1974) maps the Lebialem Highlands in the 3–4 m rainfall per annum area, but with interrogation marks, probably because there is no meteorological station in the area. Whatever the rainfall level is in the general area, precipitation is likely to higher in the immediate vicinity of the steep scarp slope since this will intercept the south-westerly moisture laden winds of the wet season, while shielding the vegetation from the arid winds of the dry season north-easterly harmattan.

## REFERENCES

Courade, G. (1974).  Commentaire des cartes. Atlas Régional. Ouest I. ORSTOM, Yaoundé.

Hawkins, P. & Brunt, M. (1965). *The Soils and Ecology of West Cameroon*. 2 Vols. FAO, Rome. 516 pp, numerous plates and maps.

# THREATS TO THE LEBIALEM HIGHLANDS

Barthélemy Tchiengué

IRAD-National Herbarium of Cameroon, BP 1601, Yaoundé, Cameroon

The Lebialem Highlands concerned with in this study are part of the Bamboutos Mountains. In this region, the main activity is agriculture. Following the same track from Baranka to the Chief's Palace at Fosimondi as Letouzey did in 1974 we discovered that forests on the slopes had been replaced by farms of food crops such as Irish potatoes, carrots, beans, maize and cabbages. The same farms are cultivated every year and fertilisers are used to improve the soil.

The montane forest had disappeared due to human activities. It is only on very steep slopes and in gullies where access is very difficult to impossible that patches of that forest survives. The steep slopes are very fragile and there had been many cases of death due to landslides in that area. This happens when it rains. It is incredible to see people farming on some of those steep slopes because it is very risky.

Our first botanical expedition in the area was in April 2004. We collected at high altitude moving downwards to Bechati but staying at Fosimondi Palace where we were provided with accommodation. During the next trip in 2005, we decided to set a camp in the forest in order to collect more extensively. Our Expedition descended to around 1300–1400 m altitude. The next time we arrived in 2006, we were unpleasantly surprised to find farms of banana and cocoyam where we had collected less than a year before. It was beyond and below these large farms that collecting was carried out in 2006. We came across *Argocoffeopsis fosimondi* here. This shows how encroachment is the main threat of the Lebialem Highland forest. By today, the area where *Argocoffeopsis fosimondi* was collected has been converted from forest to farms. Other important plants for conservation collected in that area comprised *Rhaptopetalum geophylax*, *Plectranthus insignis*, *Dischistocalyx thunbergiiflorus*.

In the Bechati-Besali area, the altitude is low. The vegetation is a made up of farms and forest. However, on the mountain slopes, human pressure is minor compared to what is found higher around Fosimondi. The main activity in the forest on these slopes is the collecting of bunches of palm oil. Since there are a lot of palm oil trees, extraction of palm oil is highly practiced. The area falls within what Letouzey (1985) called natural (?) palm oil plantations or the "palm belt" by Sanderson (1936).

## REFERENCES

Letouzey, R. (1985). Notice De La Carte Phytogéographique Du Cameroun Au 1 : 500000. : *Domaine De La Forêt Dense Humide Toujours Verte*. Institut de Recherche Agricole pour le Développement — Institut de la Carte Internationale De La Végétation. pp 95–142.

Sanderson, I. T. (1936). Amphibians Of The Mamfe Division, Cameroon. *Proceed. Zool. Soc.* London. pp. 165–208.

# THE EVOLUTION OF THIS CHECKLIST

Martin Cheek

Herbarium, Royal Botanic Gardens, Kew, Richmond, Surrey, TW9 3AE, UK

In 2004 John De Marco, former co-manager of the Bamenda Highlands Forest Project suggested to us that we consider extending out surveys of natural habitat in Cameroon to include the Lebialem Highlands. He put us in contact with Louis Nkembi of ERuDeF, the conservation NGO championing the protection of the Lebialem Forests. Thereafter we made a short reconnaissance visit to Fosimondi from our temporary base at Bali Ngemba in April 2004. At that time, we were concluding production of conservation checklists for the Bali Ngemba Forest Reserve (Harvey *et al.* 2004) and for the Kupe, Mwanenguba and Bakossi Mountains (Cheek *et al.* 2004). We were seeking to develop further such outputs with the support of a future Darwin Initiative project if there was sufficient support in Cameroon for this at the level of local action groups who were protecting forest of potentially high botanical importance. ERuDeF and the Lebialem Highlands fulfilled these criteria. Also the Lebialem Highlands, practically unknown to botanical science, had the potential to fill the gap in knowledge that existed in the c. 110 km length of the Cameroon Highlands between the Kupe-Bakossi and Bali Ngemba forest areas. While other forest areas exist within this range, it is only the Lebialem Highlands, so far as we know, which benefit from the efforts of a conservation NGO, ERuDeF and so which have we believe, the best prospects for remaining intact in future. Therefore, in our proposal to the Darwin Initiative of the UK Government for a new project, made in 2005 we included the Lebialem Highlands. Before the grant was made in mid 2006 we had already committed to two further brief surveys at Lebialem in April 2005 and Feb. 2006. Again, based at Fosimondi, with funds granted through the Overseas Fieldwok Committee of RBG Kew. Our most recent visit, based at Bechati, was made in Sept. 2006, shortly after the inaugural workshop for the new 'Red Data Plants, Cameroon' project (2006–2010) within which provision had been made for production of a book, the conservation checklist of the plants of the Lebialem Highlands. All four of our surveys were led or co-led by Dr Barthélemy Tchiengué of the National Herbarium of Cameroon. After some delays in obtaining permits to send duplicates of the specimens to RBG Kew, we were able to begin identification of the specimens, much of this work being done by Dr Tchiengué in 2008–2009, during a study visit to London from the National Herbarium of Cameroon, also sponsored through RBG Kew.

Introductory chapters for the book were completed in early 2010 allowing publication shortly thereafter.

## REFERENCES

Cheek, M., Pollard, B.J, Darbyshire, I., Onana, J.-M. & Wild, C. (eds) (2004). *The Plants of Kupe, Mwanenguba and the Bakossi Mts, Cameroon: A Conservation Checklist.* Royal Botanic Gardens, Kew, UK. iv + 508 pp.

Harvey, Y., Pollard, B.J., Darbyshire, I., Onana, J.-M. & Cheek, M. (eds) (2004). *The Plants of Bali Ngemba Forest Reserve, Cameroon: A Conservation Checklist.* Royal Botanic Gardens, Kew, UK. iv + 154 pp.

# HISTORY OF BOTANICAL EXPLORATION OF THE LEBIALEM HIGHLANDS

Martin Cheek, Iain Darbyshire, Laura Pearce & Xander van der Burgt
Herbarium, Royal Botanic Gardens, Kew, Richmond, Surrey, TW9 3AE, UK

The Lebialem Highlands of SW Region Cameroon form the steep western flank of the Bamboutos Mts and Bamiléké Plateau (W Region) within the Cameroon Highland line. The latter densely populated areas are almost completely denuded of their original natural vegetation, which would have been mainly forest. They were also relatively well studied in the latter decades of the twentieth century by French botanists such as Letouzey and Villiers, and also by Meurillon and his students, based at the University of Dschang at the southernmost extremity of the Plateau. In contrast the steep, densely forested slopes of the Lebialem Highlands remained unvisited by botanists (with the possible exception of Conrau) until 2003 when Terence Atem of ERuDeF recorded several rare plant species (Nkembi & Atem 2003). In 2004–2006 a series of surveys by the national Herbarium of Cameroon and RBG Kew, supported by ERuDeF, culminated in the production of this book. It is to be hoped that this is only the beginning of botanical exploration of the Lebialem Highlands since plant sampling levels remain low, only 548 being available to date, whereas several thousands are needed to have a reasonably complete picture of the species and vegetation types present. It is to be hoped that young Cameroonian botanists such as Robin Achah (see below) will build on the efforts documented here and surpass the efforts of her teacher, Dr Barthélemy Tchiengué of the National Herbarium of Cameroon who has collected the lion's share of specimens at Lebialem to date. Of the 548 specimen records referred to above, no less than 370 are under his numbers, and he has led, or co-led, no less than three of the four National Herbarium-RBG Kew survey visits to the Lebialem Highlands.

This chapter begins with biographic notes on the botanical specimen collectors within our study area, and concludes with excerpts from the expedition reports from the series of IRAD-National Herbarium of Cameroon-RBG, Kew rapid surveys to Lebialem in 2004–2006. Since Letouzey, Villiers, and also Meurillon appear not to have collected in our Lebialem Highlands study area (Fig. 2), their biographical details have not been included below. However some specimens collected by Letouzey have been included in the checklist since they occur in the montane grasslands close to our study area.

## G. CONRAU (–1899)

In 1898 and 1899 before his death later that year, the German botanist Conrau, engaged with the Victoria Plantation Company, collected about 200 specimens (Hepper & Neate 1971) in the northern part of present day SW Region Cameroon and adjoining NW Region. One of his collection sites, "Bangwa" or "Bangwe" is believed to be the settlement Bangwa, near Fontem, also near the western edge of the Bamiléké Plateau but about 15 km South of our study area. There is the possibility also that Bangwa is an orthographic variant of Banwa, referring to the tribal area, of which our study area is part. Many of the plant specimens collected at Bangwa by Conrau were new to science and several were named after him, such as *Magnistipula conrauana* Engl. and *Garcinia conrauana* Engl. Both these are species of submontane forest, the first occurring also in our study area. Thus it is possible that Conrau was the first to explore the western scarp of the Bamiléké Plateau in this general area. It could be held that he was the first botanist to study the Lebialem Highlands if these are taken to extend the whole of the North–South length of the Lebialem Division along the scarp slope mentioned. It would be a useful endeavour to list all the Conrau specimens and their identifications by combing the German botanical literature of the period. It would then be possible to compare Conrau's species records for Banwa with our own for Fosimondi-Bechati. Sadly Conrau seems to have collected solely or mainly unicates, as did most German botanists in former German Kamerun in that epoch (Zenker was an exception). These were nearly all destroyed at the great Berlin herbarium due to allied bombing in 1943. In the absence of these type specimens the correct application of the German scientific names linked with them to modern collections is often very difficult since the original descriptions are frequently too imprecise to be of assistance.

## B. TCHIENGUÉ (1967–)

Barthélemy Tchiengué is a Cameroonian botanist born on August 11, 1967 at Penja in Cameroon. He attended the primary school at the Holy Luke Catholic School of Penja where he got his First School Leaving Certificate in 1979. He then continued to secondary and high school both in Loum and Bafang. In 1988, he graduated the "Baccalauréat" (equivalent to Advanced Level in the Anglo Saxon system) in Biology. This certificate opened the doors of the University of Yaoundé to him, where he did Biology. After two years of general studies in Biology, he specialised in Botany and in 1991 he got his BSc in Botany. Because he did not repeat any class at the university and got his BSc with good grade, he was selected in "Maitrîse" which is a starting point of research studies and in 1993, he defended a thesis on the vegetation of Nkolobot Mountain (1050 m) in the Yaoundé area. He often organised personal field trips in order to improve his knowledge on plant identification. That is why he was quickly involved in practical and tutorial work at the University. He had the task of conducting many expeditions with students of lower level. In the scope of his doctoral degree, he carried out research on the vegetation of Mt Kupe which he defended in 2004 at the University of Yaoundé I. It was at Mt Kupe that he first encountered RBG Kew – IRAD-National Herbarium of Cameroon field survey teams in 1998 and this was the beginning of long relationship of collaboration.

Barthélemy began as a researcher at the National Herbarium of Cameroon in 2002. Because he has a good reputation as an enthusiastic fieldworker, a record of botanical specimen collecting and a mastery of the flora of West and Central Africa, he has been involved in several international botanical expeditions both in Cameroon and elsewhere, e.g. Guinea-Conakry. He is presently researching for a further PhD via the University of Frankfurt on the topic of secondary succession after shifting cultivation under a precipitation gradient. This work is funded by the German Ministry of Foreign Affairs (BMZ). Barthélemy is married and father of one.

## JAAP VAN DER WAARDE (1965–)

Jaap van der Waarde and his wife Verina Ingram (1967-) were based in Bamenda for several years with their two daughters before moving to Yaoundé. In 2010 they returned to Europe, Verina doing her PhD studies at Univ. Amsterdam, while Jaap begins in April 2010 as Manager Forest Program for WWF-Netherlands and has charge of their Central Africa forest programme. Keen naturalists, they have made many excursions to study plants and animals in different parts of Cameroon. For example, Verina Ingram joined botanical surveys to Dom (see Cheek *et al.* 2010). On 27 and 28 October 2004, van der Waarde collected and photographed plants while joining an ERuDeF gorilla tracking team based at Fosimondi. The specimens resulting from this visit have not yet reached RBG Kew for final identification. However, his beautifully produced photographic report (van der Waarde 2004), in which each of the 21 specimens he collected (van der Waarde 1–21) is illustrated, allows most of these to be fairly confidently identified to species so they are included in our checklist. His visit is of interest since, so far, his is the only botanical survey work conducted at Fosimondi at this season. In addition he is the only collector to date ever to have descended from Fosimondi through the complete submontane altitudinal range, reaching as low as 800 m alt. Sadly it was not possible for him to do very much collecting in the lower submontane altidudinal belt which remains largely unknown to botanical science. His photographic records are of interest since a high proportion are of species otherwise unrecorded from the Lebialem Highlands. His are also the first records of the refuge indicator and IUCN Vulnerable *Begonia pseudoviola* Gilg, and for example, of Asparagaceae, Gentianaceae and *Rhabdotosperma*. It is to be regretted that he could not make further collections on this visit due to logistical constraints, and that, following a very unpleasant accident with a specimen drier, he has ceased to make further plant specimens.

## ROBIN ACHAH (1983–)

Native of Batibo in the North West Region of Cameroon, Robin Achah was born on the 8th of October 1983. Educated as a Botanist at the University of Buea, she got attached to ERuDeF after her BSc. With support from RBG Kew, she continued botanical collection at the foot of the scarp, Bechati in the Lebialem Highlands initially (22 March 2007) under the tutelage of Barthélemy Tchiengué of the National Herbarium of Cameroon. During 2007, assisted by Jacob Nong she made two further botanical visits to Bechati, these being 20–24 April inclusive and 14–19 May inclusive. During these visits a highly creditable 191 specimen numbers (Achah Robin 1–191) were made (Achah Robin field note books 1 and 2). The resultant specimens suffered somewhat from slow drying and are housed at the National Herbarium of Cameroon, Yaoundé from whence they will be sent to RBG Kew for final identification. It is likely that they will then add many more species to the existing checklist. Robin currently lives in Cambridge U.K. with her husband Daniel Pouakouyou.

## NATIONAL HERBARIUM CAMEROON-RBG, KEW BOTANICAL INVENTORIES OF THE LEBIALEM HIGHLANDS (FOSIMONDI-BECHATI-BESALI) 2004–2006

In April 2004, April 2005 and February 2006, RBG Kew in collaboration with the National Herbarium of Cameroon and the support of ERuDeF mounted three short botanical surveys of the Lebialem Highlands from Fosimondi at the top of the scarp, approached via Dschang and Baranka. In September 2006 a fourth expedition was mounted, to the foot of Highlands, at Bechati, approached via Dschang and Fontem/Menji. The following details are extracted from the relevant expedition reports. Edits are indicated in [square brackets]. The reports vary greatly in their information content and style, depending on their authors, but have been only minimally edited so as to capture the maximum amount of the data they contain.

# LEBIALEM HIGHLANDS SURVEY 1: 16TH–18TH APRIL 2004 AT FOSIMONDI

(extracted from the report by Iain Darbyshire, Nina Rønsted & Terence Atem)
(as part of Earthwatch Institute, Cameroon Rainforests Team III).

RBG Kew, together with their partner institution in Cameroon, IRAD-National Herbarium of Cameroon (HNC) were requested by ERuDeF, and advised by John De Marco, the former head of the Bamenda Highlands Forest Project, to carry out a preliminary plant survey of this forest region in order to support the work on the forest fauna and its importance for conservation. With this in mind, Terence Atem of ERuDeF joined Earthwatch Team III, base at the Bali Ngemba Forest Reserve, with the intention of both participating in the fieldwork for its duration and arranging with Iain Darbyshire, the Earthwatch Principal Investigator, a three-day botanical investigation based at the village of Fosimondi.

Of particular botanical interest is the fact that this site is located between the Bamenda Highlands to the North and the Kupe-Bakossi-Mwanenguba Mts to the South and may therefore provide evidence to increase our understanding of the floristic links and differences between these montane regions.

With this in mind, the principal aim of this pilot study was to collect herbarium specimens of interesting plant taxa found within the submontane forest around Fosimondi, for return to HNC and RBG Kew for identification in order to provide preliminary conclusions on the botanical importance of the site, including whether it contains Red Data taxa, and therefore of high conservation significance, and to identify whether the area is worthy of further investigation through a full Kew-Earthwatch expedition in the near future.

Due to time constraints, the flora of the primary forest was not investigated during the current expedition.

## Participants:

ERuDeF: Terence Atem (TA)

RBG Kew: Nina Rønsted ((NR — Postdoctoral Student,, working on the genus *Ficus* (Moraceae)).

HNC: Barthélemy Tchiengué (BT — researcher), Nicole Guedje (NG — researcher).

Earthwatch volunteer: Adam Surgenor (AS — UK)

Apicultural and Nature Conservation Organisation (ANCO): Terence Suwinyi (TS), Kenneth Tah (KT)

Limbe Botanic Garden: Enow Kenneth (EK)

Bamenda Highlands Forest Project: Ndamnsa Nginyu (NN — mechanic and landrover driver).

## Expedition programme:

*Friday 16th April*

The expedition took off from Mantum village, near Bali, North West Province, at 7am, meeting NN at the Bamenda Highlands Forest Project HQ, where we transferred to John De Marco's vehicle, borrowed for this expedition. The team arrived after three hours in Mbouda, the nearest large town before the destination. Due to a Ministry visit, the town was closed and the team continued to a village (name unknown) with a market, where food and provisions for the expedition were purchased. The planned route to Fosimondi was unfortunately closed due to erosion. An alternative route followed dirt tracks for a couple of hours before entering Fosimondi village. The team was invited to stay in the house where the son of the Chief lives, which contains a number of bedrooms, running water, and large living room. Arriving in the early afternoon, the team wanted to get into the forest as soon as possible. Several villagers wanted to guide us into their forest, so we ended up splitting the team into two groups to survey patches of the submontane forest for two hours before dark. Collections here included *Anthocleista scandens*, *Heterosamara cabrae*, *Cassipourea* sp. and several species of *Begonia* inside the forest patches and a 2 m tall specimen of *Eulophia horsfallii* growing in open disturbed forest.

On our return to the camp, a formal meeting with the village was held to inform them of the reasons for our presence and our possible relation to the government etc. TA was particularly important in leading the negotiation process. The conclusion was that the villagers were eager to help, but that communication should go through the local forest council, so that the council could decide who should be involved and how it could assist us. The council would also distribute any benefits, such as payment to guides.

All material collected that day was sorted and pressed during the evening.

*Saturday 17th April*

The team set out around 7.30 with two guides and the aim of reaching the primary forest to the west of, and at a lower altitude than, Fosimondi. The team was slowed down by collecting along the path down the mountain side and by lunchtime we had two full sacs of material for pressing and were still far from reaching the primary forest.

Collections included *Beilschmiedia* sp. (in bud), *Magnistipula* sp. (flowering), *Dichapetalum* sp. (plate 6), *Pseudagrostistachys africana* subsp. *africana* and *Oncoba ovalis* (in fruit). A *Cyathea* sp. was also observed, but not collected due to time constraints.

Part of the team went back to the village to process the plant material whilst the remainder continued collecting for one to two hours, down to 1400 m alt. The return trip took approximately two hours and it should therefore be possible to reach the primary forest from Fosimondi if collecting along the way is avoided.

*Sunday 18th April*

Two hours collecting was carried out near Fosimondi village, one team on the slopes and one near a bat cave. The team left Fosimondi before lunchtime, for the return journey to Mantum, arriving there at 5pm, as NN had to return to Bamenda for the evening.

**Preliminary results and discussion:**

The preliminary survey of the submontane forests around Fosimondi included collection of 141 specimens, the majority of which were at least identified to family, many being identified to genus and some to species. At the time of writing this report, the specimens are awaiting export to Kew, thus only field determinations are to hand; much firmer conclusions will be reached following full identification of the specimens at RBG Kew and at HNC.

The collections made represent at least 48 plant families, with a further 6 specimens undetermined to family in the field (see appendix 1).

## INDICATORS OF GOOD QUALITY SUBMONTANE FOREST

Several of the taxa collected during the preliminary survey of the submontane forest in the vicinity of Fosimondi, although not in themselves rare or threatened taxa, are of ecological and conservation importance as they are strong indicators of the presence of undisturbed submontane forest stands, notably:

*Shirakiopsis elliptica* (Euphorbiaceae)

*Syzygium* × *staudtii* (Myrtaceae)

*Garcinia smeathmannii* (Guttiferae)

*Pittosporum viridiflorum* (Pittosporaceae)

*Carapa* × *procera* (Meliaceae)

*Heterosamara cabrae* (Polygalaceae)

*Xymalos monospora* (Monimiaceae)

*Cuviera longiflora* (Rubiaceae)

In addition, although not recorded during our survey, presumably because it was not producing flowers or fruits at the time of our visit, *Prunus africana* (Rosaceae) was noted in large numbers by a group surveying birds in the forest in 2002–2003 (Nkembi, 2003); this is a particularly good indicator of high quality submontane forest and is threatened by over-exploitation for timber.

### Red Data species and Red Data candidates

Several Red Data species, listed by the IUCN (2003), or Red Data Candidates proposed by Cheek *et al.* (in prep.)[Cheek *et al.* 2004) and Harvey *et al.* (in prep.)[Harvey *et al.* 2004], were recorded during the preliminary survey (NT: near threatened, VU: vulnerable, EN: endangered):

*Impatiens sakeriana* (Balsaminaceae) (Candidate: VU)

*Lobelia columnaris* (Campanulaceae) (Listed: NT)

*Pseudagrostistachys africana* var. *africana* (Euphorbiaceae) (Candidate: VU)

*Oncoba ovalis* (Flacourtiaceae) (Candidate: NT)

*Allanblackia gabonensis* (Guttiferae) (Candidate: VU)

*Anthocleista scandens* (Loganiaceae) (Candidate: VU)

*Syzygium staudtii* (Myrtaceae) (Candidate: NT)

*Aërangis gravenreuthii* (Orchidaceae) (Candidate: NT)

*Ancistrorhynchus serratus* (Orchidaceae) (Candidate: NT)

*Bulbophyllum nigericum* (Orchidaceae) (Candidate: VU)

*Diaphananthe bueae* (Orchidaceae) (Listed: EN)

*Polystachya bicalcarata* (Orchidaceae) (Candidate: VU)

*Pittosporum viridiflorum "mannii"* (Pittosporaceae) (Candidate: NT)

*Heterosamara cabrae* (Polygalaceae) (Candidate: NT)

In addition, Nkembi (2003) reports *Schefflera mannii* (Candidate: VU) [not confirmed and possibly an error for *S. hierniana*], and *Prunus africana* (Candidate: NT ) from Fosimondi, and reports

*Chassalia laikomensis* from two sites in the "Lebialem Highlands", one of which is believed to be Fosimondi (Terence Atem, pers. comm.); if confirmed, this is a Red Data-listed species believed to be Critically Endangered and thus of high conservation significance.

It is likely that full identification of the specimens will produce further Red Data candidates. Of those awaiting identification, of particular interest are:

*Magnistipula* sp.: several species of this genus are highly restricted in range. Our specimen may refer to *M. conrauana*, an endangered taxon and Red Data candidate known only from west Cameroon.

*Beilschmiedia* sp.: this poorly known genus contains many highly localised species.

*Plectranthus* sp.: the two collections at Fosimondi were from rocks by a waterfall, ideal habitat for the recently described *P. cataractarum* which is Red Data-listed as vulnerable.

*Rinorea* sp.: the two collections perhaps represent the submontane species *R. preussii*, a Red Data candidate as near-threatened.

*Begonia* spp.: many begonias are highly restricted in distribution and threatened by habitat loss.

The significant number of Red Data species/candidates and other potentially threatened taxa recorded in such a brief collecting period highlights the conservation importance of this site. Furthermore, this only considers the submontane element of the forest; the primary lowland forest at this site is likely to produce a range of other threatened taxa.

## Species of phytogeographical significance

Exploration at Fosimondi should provide significance increases in our knowledge of species ranges and may provide clues as to the reasons for range limits in the western Cameroon uplands. From the initial survey, several interesting taxa have been noted:

### *Lobelia columnaris* and *Diaphananthe bueae*
These species were previously recorded only in the Bamenda Highlands and on Mt Cameroon (Cheek *et al.*, 2001), being absent from the Kupe-Bakossi-Mwanenguba Mts and not previously recorded in the Bamboutos. These records both extend their known ranges and also raise further the question of their absence from Kupe-Bakossi-Mwanenguba.

### *Heterosamara cabrae*
This species has only been recorded as far North as Mwanenguba previously, being unknown from the Bamenda Highlands. The two records from Fosimondi therefore extend its range and raise the question of its absence from submontane forest in the Bamenda Highlands e.g. at Bali Ngemba Forest Reserve.

In addition, several taxa recorded here but highly localised in global terms, being restricted to the west Cameroon mountain chain, are recorded in both the Bamenda Highlands and Kupe-Bakossi-Mwanenguba, such as *Allanblackia gabonensis*, *Anthocleista scandens*, *Impatiens sakeriana* and *Oncoba ovalis*.

Initial findings therefore suggest that the Fosimondi submontane forest contains a cross-over of taxa from both the Bamenda Highlands and Kupe-Bakossi-Mwanenguba, and may be a significant site for recording range limits of rare taxa and the reasons behind these. However, further collections are needed to consolidate this idea.

**THREATS TO THE FOREST**

Initial investigations of the forest at Fosimondi indicate that, although disturbance in the submontane forest is relatively high, with widespread agriculture and farmbush, some of the remaining forest stands are relatively undisturbed and thus contain a rich flora. Furthermore, the local community appear dedicated to the protection and sustainable use of their forest resources, thus positive conclusions should be drawn as to the future of this site. However, a formal recognition of community forest status will further promote forest protection and it is therefore important to gather data on the flora and fauna of the forest and its importance in terms of both conservation and the future prosperity of the community.

Although not observed directly, the lowland and mid-elevation forests east of Bechati and Besali are said to be extensive and with only limited disturbance (Terence Atem, *pers. comm.*). The presence of large primates here should play a major role in promoting conservation, but gathering of data on the floristic diversity would strengthen the case for formal conservation status.

**CONCLUSIONS: PROSPECTS FOR FUTURE BOTANICAL SURVEY WORK IN THE BECHATI-FOSIMONDI-BESALI FOREST**

The Bechati-Fosimondi-Besali forest includes a transect from disturbed submontane cloud forest to primary lowland rainforest. The preliminary survey of the flora was restricted to the submontane forest, but has already revealed this area to be of high botanical interest, with the presence of a good number of Red Data species.

Access to the forest is good, as research teams can both be based in the upper village of Fosimondi, with access to the submontane forest, and in Bechati village at the edge of the primary lowland forest. If this proves difficult, the primary forest could be reached from Fosimondi, though this would involve a daily hike of more than 2 hours each way, avoiding distraction from the forest patches along the way. One difficulty noted is the current state of the roads in and out of the region, which are believed to become difficult to pass during the season of heavier rains.

A further advantage to the region is the huge interest from the local villagers in the conservation of their forest, which allows for effective collaboration during a longer-term survey of the forest as well as enhancing the potential for setting up a successful conservation management programme for the future, either as a community forest or as a reserve. Communities in Cameroon have only recently had the opportunity to gain legal entitlement to forest resources — since the passing of a new Forestry Law in 1994 — and this seems to be the way forward to achieve local respect for and responsible use of the forest.

Finally the presence of both large primates and endangered bird species strengthens the potential need for conservation of the Bechati-Fosimondi-Besali forest.

It is therefore recommended that collaborative fieldwork between RBG Kew, IRAD-National Herbarium of Cameroon and ERuDeF, with the support of the Earthwatch institute, be carried out, initially for a two-week period in the spring of 2005 in order to begin gathering botanical data which will contribute to the production of a "Conservation Checklist on the plants of the Bechati-Fosimondi-Besali Forest". It is perceived that such a publication could provide data crucial to the formal recognition of this site as an area of high conservation significance.

# ACKNOWLEDGEMENTS

We are indebted to the people of Fosimondi, particularly the members of the forest council and our guides, who made fieldwork in their forest possible. We also thank the IRAD-National Herbarium of Cameroon for their contribution to the expedition and for their continuing support and collaboration with RBG Kew. We thank all those who collected in the field with us, and Ndamnsa Nginyu who so expertly negotiated the difficult driving conditions. Finally, we thank the Earthwatch Institute and the Overseas Fieldwork Committee of RBG Kew for providing funds to allow this expedition to take place.

**Note**: Not included here: Appendix 1. List of plants collected on the RBG Kew-HNC-Earthwatch expedition to Fosimondi, in association with ERuDeF, 16th–18th April 2004 (orchid taxa recorded under the numbers of Simo Placide, University of Yaoundé I). [Briefly numbers recorded were as follows: Tchiengué 1903–1971 (70 numbers); Simo, P. 218–237 (30 numbers); Tah, Kenneth, 286–324 (41 numbers); Rønsted 237–240 (41 numbers). Total 274 specimen records.]

## LEBIALEM HIGHLANDS SURVEY 2: 26TH– 30TH APRIL 2005 AT FOSIMONDI
(extracted from the report by Laura Pearce)
(funded by OFC committee, RBG Kew).

**Participants**:

ERuDeF: Dennis Ndeloh (DN)

RBG Kew: Laura Pearce (LP) and Aline Horwath, volunteer (AH).

HNC: Barthélemy Tchiengué (TB — researcher).

[119 specimen records were made: Tchiengué 2163–2288]

**Expedition programme:**

*Tuesday 26th April*

At 7.30am TB, TA, AH & LP depart for Fosimondi (Western Bamboutos). Bus from Loum arrives in Dschang at 4pm. Meet DN, a collegue of TA's from ERuDeF (the Environment and Rural Development Foundation) and buy provisions for the coming week. TA is unable to continue with the team as he has to travel to Yaoundé for a meeting the next day. DN joins the team in his place. Leave Dschang at 7pm and arrive in Mbelenka [also Baranka], the closest market town to Fosimondi at 9pm. As it is dark and there are no vehicles going down to Fosimondi in the evening, we book a room in a hostel for the night.

*Wednesday 27th April*

7am DN stays in Mbelenka to arrange transport for the luggage. TB, AH & LP walk along road leading from Mbelenka to Fosimondi and collect, looking out for *Ternstroemia* (last seen in 1974) in particular. The vegetation is mostly very degraded. There are very few woodland patches left and those that are, are mostly inaccessible due to the steep terrain. No *Ternstroemia* are found. Most collections made are by the roadside and in the montane grassland. One or two accessible thickets were explored. Arrive at Fosimondi at approx. 4pm. Meet Fonang Festus, President of the community forest development, who leads the team to the Fons palace. The Fon is away for the week and so unable to meet the team. Press specimens till dark — no electricity. Dennis kindly cooks for the team.

*Thursday 28th April*

Meet guides (Tadjong Thomas and Mepielebah Samson) and arrange to take one of them for the day due to the small numbers on the team. Mepielebah Samson decides to accompany team today. Depart at 7.30am taking the path from Fosimondi to the Besali-Bechati forest. Steep path down through secondary submontane forest and clearings with crops descending to 1340 m. Pressed and lunched at margin of primary forest. In afternoon followed path through primary forest collecting on the way. Took steep path back up to the village. Pressed specimens till 8.30 using lamps to light our work. Evening meal prepared by local lady. Previous day's specimens packed in alcohol.

*Friday 29th April*

At 6.30am depart for forest, heading North on the recommendation of the guide Tadjong Thomas. Walk though grassland, crops and finally into fragmented forest where follow dry stream-bed down. Find that forest of original destination has been burnt for crops. Head back to Fosimondi with detour through forest fragment. From Fosimondi take path to Bechati passing through submontane forest with *Allanblackia gabonensis* (Pellegr.) Bamps and *Santiria trimera* (Oliver) Aubrév. At approx. 3.30 collecting hampered by heavy rain so walk back to the palace. Press from 4.30 to 8.30.

Previous day's specimens packed in alcohol.

*Saturday 30th April*

At 7am walk up to the village to meet truck that is to transport team to Mbelenka. Truck is full of Irish potatoes. AH & LP travel in cabin with driver and young son. TB and DN bravely climb on top of load with baggage. Heart-stopping drive up steep road to summit with many stops to shovel sand onto road for the Toyota to grip on. Bus to Dschang leaves at approx. 11am. At Dschang say farewell to DN and catch bus that takes new faster road to Loum Junction. Back at Nyasoso just before nightfall.

**Results**

163 collections were made in the range TB 2134–2295 some of which have not been included in the checklist as they fall under the gazette of Mbelenka. The most interesting and significant collections made under the Fosimondi gazette during this expedition are listed below.

*27th April*

Red Data taxa:

*Schefflera hierniana* Harms (Araliaceae) IUCN: EN

*Raphidiocystis mannii* Hook.f. (Cucurbitaceae) IUCN: NT

*Gladiolus aequinoctialis* Herb. (Iridaceae) IUCN: NT

*Plectranthus punctatus* L'Hér. subsp. *lanatus* J.K.Morton (Labiatae) IUCN: VU

*28th April*

Significant collections:

The first collections of *Argocoffeopsis fosimondi* Tchiengué & Cheek (Rubiaceae) IUCN: CR

Red Data taxa:

*Uvariodendron connivens* (Benth.) R.E.Fr. (Annonaceae) IUCN: NT

*Impatiens letouzeyi* Grey-Wilson (Balsaminaceae) IUCN: EN

*Begonia preussii* Warb. (Begoniaceae) IUCN: VU

*Aneilema dispermum* Brenan (Commelinaceae) IUCN: NT

*Macaranga occidentalis* (Müll.Arg.) Müll.Arg. (Euphorbiaceae) IUCN: NT

*Plectranthus insignis* Hook.f. (Labiatae) IUCN: NT

*Heckeldora ledermannii* (Harms) J.J. de Wilde (Meliaceae) IUCN: EN

*Jasminum preussii* Engl. & Knobl. (Oleaceae) IUCN: NT

*Psychotria martinetugei* Cheek (Rubiaceae) IUCN: NT

*Sherbournia zenkeri* Hua (Rubiaceae) IUCN: NT

*Tarenna vignei* Hutch. & Dalziel var. *subglabra* Keay (Rubiaceae) IUCN: NT

*Leptonychia kamerunensis* Engl. & K.Krause (Sterculiaceae) IUCN: EN

*Urera gravenreuthii* Engl. (Urticaceae) IUCN: NT

Interesting / as yet unplaced taxa:

*Amorphophallus* sp. of Fosimondi (Araceae)

*Ficus* sp. of Fosimondi (Moraceae)

*Cyrtorchis* sp. of Fosimondi (Orchidaceae)

*Mussaenda* sp. 1 of Fosimondi (Rubiaceae)

*Aframomum* sp. 3 of Fosimondi (Zingiberaceae)

*29th April*

Red Data taxa:

*Begonia oxyanthera* Warb. (Begoniaceae) IUCN: VU

*Eragrostis camerunensis* W.D.Clayton (Gramineae) IUCN: VU

*Psorospermum aurantiacum* Engl. (Guttiferae) IUCN: VU

*Zenkerella citrina* Taub. (Leguminosae–Caesalpinioideae) IUCN: NT

*Quassia sanguinea* Cheek & Jongkind (Simaroubaceae) IUCN: VU

*Cola anomala* K.Schum. (Sterculiaceae) IUCN: NT

Interesting / as yet unplaced taxa:

*Tabernaemontana* sp. of Bali Ngemba (Apocynaceae)

*Aframomum* sp. 1 of Fosimondi (Zingiberaceae)

## ACKNOWLEDGEMENTS

We could not have completed this work without the help and kindness of the people of Fosimondi, Dennis Ndeloh and Terence Atem of ERuDeF and the staff of the IRAD-National Herbarium of Cameroon for their support in Cameroon. We also thank the Overseas Fieldwork Committee of RBG Kew for funding the expedition.

## LEBIALEM HIGHLANDS SURVEY 3: FEBRUARY 2006 AT FOSIMONDI
(funded by OFC committee, RBG Kew).

**Participants:**

ERuDeF: Dennis Ndeloh

RBG Kew: Aline Horwath, volunteer (AH).

HNC: Barthélemy Tchiengué (TB — researcher).

Fosimondi: Columbus

Volunteer: Bate Oben (BO)

No report has been traced for this visit but a database check shows that the specimens made were in the range *Tchiengué* 2560–2646 inclusive (86 numbers), over the 23rd, 24th and 25th February. The expedition was based at Fosimondi.

## LEBIALEM HIGHLANDS SURVEY 4: SEPTEMBER 2006 AT BECHATI
(extracted from the report by van der Burgt & Tchiengué)
(funded by Darwin Initiative Red Data Plants Cameroon project).

**Participants:**

RBG Kew: Xander van der Burgt

HNC: Barthélemy Tchiengué

Bechati & ERuDeF: Khumbah, P, Nemba Nong, J.

Arriving in Dschang from Yaoundé on the evening of Friday 22nd Sept 06 were Xander van der Burgt (XVDB, RBG Kew), and Barthélemy Tchiengué (BT, Nat. Herb. Cameroon).

*Sat. 23rd Sept 06*

The team travelled to Menji [also known as Fontem], HQ of ERuDeF and meeting Pius Awungjia who was to join the group and introduce them to the community of Bechati and local offices. Information on these introductions and meetings is contained in a report on the visit by Pius, adjoined to this one. From Menji, three motorbikes were taken to Bechati Village. The bikes were carried across the intervening rivers. The journey took about four hours, in the course of which Tchiengué's leg was scalded by the bike exhaust. It was arranged that the bikes should return the pm of the 27th to be able to transport the group to Menji on 28th.

*Sun 24th – Weds 27th Sept 06*

Sunday 24th was the first of four full days patrolling and specimen collecting by the group in the course of which numbers gathered were as follows:

X vd Burgt    850–900 (51 numbers)

B Tchiengué   2753–2842 (90 numbers)

On different days the group took three different paths with local guides and ascended the forested slope to an altitude of c. 620 m alt. The area around Bechati itself was dominated by *Elaeis guineensis*, the oil palm. Near the village there were a number of plantations of *Elaeis guineensis* and *Musa*. On the slopes above these plantations a secondary forest and shrubland was present, with scattered oil palms that were regularly tended and harvested. It did not become clear whether these oil palms were natural or planted long ago. A single large *Baillonella toxisperma* left standing so that its seeds might be harvested, was the last remnant of the original high forest. As altitude increased their numbers declined. Around 600 m alt. there were few oil palms left and a less disturbed, closed canopy forest was present.

**Habitat conservation assessment-Bechati Forest of Lebialem Highlands.**

In non-colonised areas, the forest scraps encountered contained tree species indicating high quality submontane-lowland forest for which Bechati is the northernmost scientifically recorded site. This in itself makes the forest of conservation interest. The species concerned are:

*Napoleonaea egertonii* (other locations being Mt Juahan, Oban Hills, Mts Bakossi and Kupe)

*Leonardoxa africana* (more widespread, but generally in good quality lowland forest)

*Medusandra mpomiana* (otherwise only known from Kupe-Bakossi Mts and adjoining areas) and a species endemic to Western Cameroon.

Along a mountain stream near the village, several collections of interesting trees and lianas were made, as well as the orchid *Habenaria weileriana,* growing on bare rock submerged after heavy rain.

The above identifications were made in the field. There was not enough time in Yaoundé to attempt to name most of the 141 numbers gathered at Bechati. This will have to wait until they arrive in London — permits are awaited.

However, three of the numbers were identified and proved especially interesting:

A very unusual climbing-scandent Commelinaceae: *Coleotrype laurentii*

The world's second only collection of a very rare Melastomataceae *Cincinnobotrys letouzeyi* from mossy rocks near waterfalls.

*Begonia pseudoviola,* a narrow endemic from the western Cameroon Mts. This is a yellow-flowered terrestrial *Begonia* with slightly bullate leaves, the upper surface sparely pilose, identified by Professor Sosef of Wageningen (from a digital photo).

The above indicators suggest that the Bechati area is potentially of high interest for plant conservation.

*Thurs 28th Sept 06*
Bechati – Menji, Menji – Dschang

*Fri 29th Sept 06*
Dschang – Yaoundé, National Herbarium, drying specimens.

## RED DATA PLANTS COLLECTION BECHATI SEPTEMBER 24TH–29TH 2006
### (Report by Pius Awungjia Khumbah (field staff ERuDeF))

**INTRODUCTION**

ERuDeF is a non–governmental organisation, working in Lebialem Highlands Forest project located in the western part of Mount Bamboutos. One of its main objective is the conservation of biological diversity with focus on great apes, plants and birds. ERuDeF's potential research and conservation sites are found in the villages of Bechati and Besali in Wabani subdivision and in Fosimondi village Alou subdivision. The forest in this villages happens to harbour the critically endangered gorilla, chimpanzee, other large mammals and many globally threatened birds species like the rock fowl (Grey-necked picarthates) and Bannerman's Turaco (*Touraco bannermani*) etc. The forest in the above mention villages has been suspected to equally be rich in promising plants species. Since community-based conservation of great apes and birds is going along with the conservation of forest and plants as well, it is therefore important that a detail research on red data plants of the above villages be carried out and while not for the whole division. Equally nobody has ever written any documentation about the rare plants of Lebialem. The division is thus considered to be the only division in the South West Province that nothing has ever been said about its red data plant collection.

In this light a botanical expedition was carried out in Bechati September 2006 for four days. The team that carried out this survey was made of: 01 specialist in plant collection (Mr Xander van der Burgt) from the Royal Botanic Gardens Kew in the UK, 01 leading Botanist from the National Herbarium of Cameroon (Dr Barthélemy Tchiengué) base in Yaoundé, 01 field staff from ERuDeF's great apes research and conservation program (Pius A. Khumbah) and a local guide.

**OBJECTIVES**

The objectives of this very important ongoing survey is to came out with a checklist of red data plants that all communities whose forests shall be surveyed in the Mundani area will own a copy, to identify all endangered plants of the forest that have been found in other places in Cameroon and those that have never been found any where in Cameroon apart from the Bechati area (endemic to Bechati-Cameroon). The above objectives will help ERuDeF to came up with more better managenent plans for the forest in future.

**METHODS**

Immediately we got the Bechati community, all members of the Bechati forest management community — women, men, children, hunters and other interested persons — were invited to meet us in the Fon's palace at 8.00 pm. We had a meeting with all this categories of people during which we briefed all of them on our mission to the area and what we had to do in their forest. Dr Tchiengué from the national herbarium of cameroon used the opportunity to tell the Bechati people that a similar exercise has been done on slopes as high as 1800 m in the Fosimondi area twice. We now need to know the plants of the Bechati area because it is a low-lying region and vegetation changes with increase or decrease in altitude. We made these people to understand that Lebialem division was the only division in the South West Province that nothing has ever been said about its plants diversity. We sited places like Mt Cameroon, Mt Oku, Mt Kupe and Mwanenguba along the Mt

Cameroon chain where Lebialem is also located as places that similar exercises have taken place. We also made the people to understand that we were collecting these plants for scientific research only and not for commercial purpose. The people were made to understand the whole issue of sustainability because every where people depend directly on forest, plants and animals to sustained a living. We sited some disadvantages of forest loss like, loss of flavours (spices), loss of medicinal herbs, migration of important animal and bird species that could lead to their extinction, loss of fresh water sources and soil degradation. A question was asked by one of these villagers that are we out to collect backs of trees or not? Response; the people were made to understand that our interest was more on collecting plants in their flowering and seed production stages only. The next day our team left for the forest for plant collection. Collection of red data plants was mostly done while moving along Hunters trails and searching for flowers. Our speed of movement was slow in order to maximise collection. All the plants that were collected were either in their seed production or flowering stage. A pair of sharp scissors was used in cutting and collecting the plant. Collecting bags were used in temporally storing the plants before pressing sessions. GPS coordinates were taken at every site that a collection and pressing session was done. An altimeter was used in determining the altitude of any site on which collection had been done. The highest altitude we got to was 650 m. A pocket lens was always used to better observe some plants in some cases. In every case the two specialists at least made the rest of the team to understand the class or family to which every red data plant collected belongs to. At every pressing session each red data plant collected was carefully put onto a press with the two sides of the leaves of that plant displayed. Tags carrying the number and name of the person who collected the plant were put onto each plant. After this the specimens were carefully put between two solid boards and were firmly attached using strong strings. After 24 hrs each day pressed specimens were put in solid plastic bags and four cups of alcohol were put on them to preserve them. After this, the plastic bags were firmly tied to protect the alcohol from vaporising. At this stage they were ready to be taken to the national herbarium for laboratory observation.

## RESULTS

We collected more than 150 different kinds of specimens in four days. A checklist of all red data plants collected in Bechati september 2006 is now available in the Royal Botanic Gardens Kew in London care of Dr Martin Cheek. The people of Bechati all cherish the idea of research and conservation of promising plants. They however do not link plants to totems as it's the case with great apes.

## Recommendations:

That more time be spent for collection on the many slopes in Bechati. That Bechati is practically inaccessible during the rainy season so if surveys are organised more during the dry season it will be much better.

## CONCLUSION

The number of specimens that were collected by the team within four days was encouraging. It can be concluded that the Lebialem Highlands are rich in biological diversity and will have some of the rare of plants of Cameroon.

PIUS AWUNGJIA KHUMBAH, November 2006

# REFERENCES

Cheek. M., Harvey, Y. & Onana, J.-M. (eds) (2010). *The Plants of Dom, Bamenda Highlands, Cameroon: A Conservation Checklist.* Royal Botanic Gardens, Kew, UK. iv + 162 pp.

Cheek, M., Onana, J.-M. & Pollard, B.J. (2000). *The Plants of Mount Oku and the Ijim Ridge, Cameroon, A Conservation Checklist.* Royal Botanic Gardens, Kew, UK. iv + 211 pp.

Cheek, M., Pollard, B.J, Darbyshire, I., Onana, J.-M. & Wild, C. (eds) (2004). *The Plants of Kupe, Mwanenguba and the Bakossi Mts, Cameroon: A Conservation Checklist.* Royal Botanic Gardens, Kew, UK. iv + 508 pp.

Harvey, Y., Pollard, B.J., Darbyshire, I., Onana, J.-M. & Cheek, M. (eds) (2004). *The Plants of Bali Ngemba Forest Reserve, Cameroon: A Conservation Checklist.* Royal Botanic Gardens, Kew, UK. iv + 154 pp.

Hepper, F.N. & Neate, F. (1971). *Plant collectors in West Africa: a biographical index of those who have collected herbarium material of flowering plants and ferns between Cape Verde and Lake Chad, and from the coast to the Sahara at 18°n.* International Bureau for Plant Taxonomy and Nomenclature. Utrecht: Oosthoek. xvi, 89, 8p. of plates.

IUCN (2003). *Guidelines for Application of IUCN Red List Criteria at Regional Levels: Version 3.0.* IUCN Species Survival Commission. IUCN, Gland, Switzerland and Cambridge, UK.

Nkembi, L. & Atem, T. (2003) *A Report on the Biological and Socio-economic Activities conducted by the Lebialem Highlands Forest Project, South West Province, Cameroon.* Environment and Rural Development Foundation Report.

van der Waarde, J. (2004). *Plant Collection Around Fossimondi, Bamboutos Mountains, South West Province, Cameroon, 27 and 28 October 2004.* Internal Project Report, project code 2004.022. 32 pp, November 2004.

# ENDEMIC, NEAR-ENDEMIC & NEW TAXA AT LEBIALEM HIGHLANDS

Martin Cheek

Herbarium, Royal Botanic Gardens, Kew, Richmond, Surrey, TW9 3AE, UK

Below are listed the 17 endemic or near-endemic taxa of the Lebialem Highlands. Of these, 6 are considered as strict endemics, i.e. only known from Lebialem Highlands, and 11 as near-endemics (known at one other site).

## LEBIALEM STRICT ENDEMICS

*Monanthotaxis* sp. nov. of Bechati (endemic to Lebialem, more material needed to formally name).

*Salacia* sp. aff. *nitida* of Fosimondi (endemic to Lebialem, more material needed to formally name).

*Dorstenia barteri* var. nov. of Bechati (endemic to Lebialem, not yet formally named).

*Argocoffeopsis fosimondi* (endemic to Lebialem, described in this volume).

*Chassalia* sp. 1 of Bechati (endemic to Lebialem, not yet formally named)

*Mussaenda* sp. 1 of Fosimondi (endemic to Lebialem, more material needed to formally name).

Further survey work and studies may yet discover these species at other sites outside the Lebialem Highlands but for the present they are known uniquely from the Lebialem Highlands. It is likely that further work will reveal additional endemic plant species here.

## LEBIALEM NEAR-ENDEMICS

*Cincinnobotrys letouzeyi* (also near Mamfe).

*Beilschmiedia* sp. 1 of Bali Ngemba checklist (also Bali Ngemba)

*Heckeldora ledermanii* (also Kupe-Bakossi)

*Coffea montekupensis* (also Kupe-Bakossi)

*Lasianthus* sp. 1 of Kupe checklist (also Kupe-Bakossi)

*Pentaloncha* sp. nov. of Kupe Bakossi (also Kupe-Bakossi)

*Psychotria babatwoensis* (also Bali Ngemba/Baba II)

*Psychotria* sp. B of Bali Ngemba checklist (also Bali Ngemba)

*Psychotria* sp. 9 of Kupe-Bakossi (also Kupe-Bakossi)

*Trichostachys petiolata* (also Kupe-Bakossi)

*Rhaptopetalum geophylax* (also Kupe-Bakossi)

# ADDITIONAL NEW LEBIALEM TAXA

In addition to the eleven new taxa listed above for Lebialem Highlands, a further five new species are known. These are either recently described, based on Lebialem specimens, or awaiting description. All are known from two or more locations in addition to Lebialem Highlands.

*Tabernaemontana* sp. of Bali Ngemba (also Tabenken, Dom, Kupe-Bakossi, Bali Ngemba)

*Voacanga* sp. 1 of Bali Ngemba & Kupe (also Kupe-Bakossi, Bali Ngemba)

*Myrianthus fosi* (described in this volume, also Kupe-Bakossi, Rumpi Hills, Ebo, Ndikinimeke, SW Mamfe)

*Tricalysia* sp. B aff. *ferorum* (also Kupe-Bakossi, Bali Ngemba)

*Deinbollia oreophila* (published in 2009, also Mt Cameroon, Kupe-Bakossi, Ebo)

# *ARGOCOFFEOPSIS FOSIMONDI* (RUBIACEAE) A NEW CLOUD FOREST SHRUB SPECIES FROM FOSIMONDI, CAMEROON

Martin Cheek[1] & Barthélemy Tchiengué[2]

[1]Herbarium, Royal Botanic Gardens, Kew, Richmond, Surrey, TW9 3AE, UK

[2]IRAD-National Herbarium of Cameroon, BP 1601, Yaoundé, Cameroon

**Abstract:** A new species, *Argocoffeopsis fosimondi* Tchiengué & Cheek is described from Fosimondi in the Lebialem Highlands of Cameroon. It is assessed as Critically Endangered.

This attractive new species of shrub which has good potential as an ornamental was discovered among the specimens resulting from a series of expeditions to the Lebialem Highlands led by Barthélemy Tchiengué. In the field it was assumed to be either a *Coffea* or *Tricalysia* since like those genera it has axillary fascicles of flowers, each subtended by several calyculoid bract-bracteoles. The species particularly resembles a *Coffea* since the calyx limb is reduced to an inconspicuous unlobed rim and the buds are often protected by a bead of translucent gum. It differs notably in the seeds which lack the papery endocarp and ventral groove of *Coffea*.

The affinities of this shrub are with a recently described species, also from Cameroon, *Argocoffeopsis spathulata* A.P.Davies & Sonké (2009) and with an undescribed species from Bakossi, Cameroon, treated as *Calycosiphonia* sp. A by Cheek (Cheek *et al.* 2004: 369). All three taxa are glabrous evergreen shrubs or treelets with stems that dry black and later develop a white, spongy, exfoliating epidermis. All three have large elliptic-oblong leaves and mainly 5- or 6-merous flowers with non-septate anthers borne in paired axillary fasciculate inflorescences. They differ from most *Argocoffeopsis*, which are hairy lianas with inflorescences terminal on axillary shoots with 5-merous flowers resembling instead *Calycosiphonia* Pierre ex Robbr. in habit and inflorescences, but lacking the septate anthers and 7–8-merous flowers of that genus. However intermediates occur, such as *A. lemblinii* (A.Chev.) Robbr. which is 8-merous and *A. eketensis* has 5–10 corolla lobes. Essentially *Calycosiphonia* appears to be distinguished from *Argocoffeopsis* as currently defined by only a single character, septate anthers, and so probably deserves to be synonymised. On the other hand it may be that *Argocoffeopsis* is better divided in two, by the resurrection of *Argocoffea* Lebrun. These questions are beyond the objective of this paper, which is to delimit this new taxon, justify it and publish it formally in accordance with the Code so as to draw attention to its existence and help focus attention on its conservation. Characters separating *A. spathulata* and our new species, *A. fosimondi*, are documented in Table 1 below. The description mostly follows the format of Davies & Sonké (2009). All specimens cited have been seen.

**TABLE 1.** The more significant characters distinguishing *Argocoffeopsis spathulata* and *A. fosimondi*

| | *Argocoffeopsis spathulata* | *Argocoffeopsis fosimondi* |
|---|---|---|
| leaf-blades | (8.5–)11–17 × 2.5–5.7 cm | 16–21.5 × 4.5–9.5 cm |
| leaf acumen | spatulate, apex widened | tapering to acute apex |
| secondary nerves | joining to form hooped intramarginal nerve | intramarginal hooped nerve not apparent |
| petiole length | (2–)4–7(–8) mm | (9–)11–17 mm |
| basal calyculi | 0.5–1 × 1–1.5 mm | 1.6–2 × 2–3.4 mm |
| upper calyculi | 0.7–1 × 1.5–2 mm | 2–3 × 3–4 mm |
| corolla tube length | 4.5–6(–10) mm | 8–10 mm |
| corolla lobe length | (2.5–)7–9 mm | 16–20 mm |
| anther length & apex | 2.5–3 mm, subacute | 9 mm, apiculate |
| fruit | obovoid, 12–15 × 5–10 mm | globose, c. 30 × 25 mm |

**Argocoffeopsis fosimondi** Tchiengué & Cheek **sp. nov.** ab A. spathulata A.P.Davis & Sonké acuminis foliorum angustatis acutisque (non spathulatis), fructibus subglobosis usque 2 cm diametro (non obovoideis 1.2–1.5 × 1.5–2 mm), foliis majoribus 16–21.5 × 4.5–9.5 cm (non 11–17 × 2.5–5.7 cm) distincta. Typus: Cameroon Highlands, W. Bamboutos Mts/Lebialem Highlands, Fosimondi, Ntoo Forest, fl./fr. 24 Feb. 2006, *Tchiengué* 2597 (holotypus K!; isotypi BR!, WAG!, YA!).

Shrub (1.5–)2–3 m tall, glabrous. Stem terete, 2–3 mm diam., drying matt black in first and second internodes, in the second to fourth internodes the epidermis becoming corky white, subglossy and exfoliating as in a *Chazaliella*; internodes 3–8 cm long. Leaf-blades opposite, equal, elliptic-oblong, rarely ovate-lanceolate, 16–21.5 × 4.5–9.5 cm, acumen tapering, slender, acute, 0.5–2 cm long, base acute, slightly to conspicuously asymmetric, the blade up to 8 mm longer on one side than the other, lateral nerves 6–9 on each side of the midrib, arising at 50–70 degrees from the midrib, curving upwards and becoming inconspicuous c. 3 mm from the margin, domatia absent; tertiary and quaternary veinlets visible to the naked eye but not conspicuous, reticulate, margin entire; drying dark green above, pale green below, with midrib pale to dark brown; petiole inserted at 45° from stem, flattened above, concave below (9–)1.1–1.7 × 0.1–0.15 cm, drying black. Stipule with basal cylindrical sheath c. 2 × 4 mm; distal limb absent to triangular 0–1.5 mm long with central ridge absent or present, extending into an arista 0.5–2 mm long, inner surface glabrous, lacking colleters or other hairs, bearing only deposits of exudate. Inflorescences on leafy branches (2–)3–4 internodes from apex, subtended by leaves, inserted 2–5 mm above the leaf axil; fasciculate, peduncles 1-flowered, 1–3 per axil at both axils of node; 1–3 axils fertile per stem. Calyculi 2, the uppermost slightly larger, each calyculus cylindrical-campanulate, more or less 4-lobed (2 foliar lobes and 2 stipular lobes), rachis hidden by calyculi, glabrous, colleters not seen; basal (1st) calyculus 1.4–2 × 1.9–3.4 mm, foliar lobes reduced to mucra, 0–0.3 mm long, stipular lobes absent or minute; upper (2nd) calyculus 2–3 × 2.5–4 mm, foliar lobes narrowly triangular or ligulate, 0.4–3 mm long, stipular lobes 0.04–0.5 mm long. Flowers hermaphrodite, secondary pollination mechanism present, homostylous, (5- or) 6-merous, sessile. Calyx (hypanthium) entirely contained within the upper calyculus at anthesis, c. 2 × 2 mm, glabrous; calyx limb truncate, glabrous, lacking colleters inside, barely detectable. Corolla funnel-shaped, glabrous, white; corolla lobes contorted to the left in bud; corolla tube cylindrical, widening slightly in the apical 2 mm, 8–10 mm long; corolla lobes oblong-elliptic, (12–)15–20 × 7–8 mm, apices asymmetrically apiculate, apiculus 1 mm. Anthers fixed at corolla throat, completely exserted, submedifixed; filaments 2–2.2 mm long; anther sacs 9 mm long, base rounded, apex with connective apiculate, 0.6 mm long. Disc subcylindrical, c. 0.6 × 1.3 mm, glabrous. Ovary c. 2 × 2 mm (at flowering stage), bilocular, each locule with a single downward facing anatropous, possibly unitegmic, ovule (inferred from fruit since flowering ovary not dissected); style filiform, 16 × 0.6 mm, glabrous, fleshy, 5.5 mm, bilobed, exserted. Fruits orange, globose, c. 30 × 25 mm including a broad 4 mm long stipe with accrescent calyculi; apex with disc accrescent, convex, 8 mm wide, calyx limb unlobed, forming a narrow rim around the disc; outer surface matt, subrugulose, mesocarp leathery (becoming fleshy?), 4–6 mm thick; endocarp not woody, not strongly differentiated. Seeds 1–2, flattened and orbicular in outline 12 × 12 × 3 mm (where 2 seeds per fruit) or ellipsoid 13 × 11 × 10 mm (where 1 seed per fruit). Seedcoat distinct, brown, glossy, vascularised, probably pachychalazal, c. 0.05 mm thick, hilar area extending more or less the length of the seed, 5–7 mm wide at the middle, tapering to a point at base and apex; micropyle slit-like extending from the back of the seed to about half way along the dorsal surface. Endosperm waxy, forming 98% of the seed. Embryo straight, linear, running along the axis of the seed, cotyledons 2 mm wide, flattened. Fig. 5.

**DISTRIBUTION.** Cameroon, Cameroon Highlands, known only from Fosimondi (Lebialem Highlands near W Bamboutos Mts).

**HABITAT.** Submontane (cloud) forest with *Santiria trimera* (Burseraceae), *Cola verticillata* (Sterculiaceae), *Macaranga occidentalis* (Euphorbiaceae), *Chassalia laikomensis*, *Pauridiantha paucinervis* (both Rubiaceae), *Carapa procera* (Meliaceae). 1330–1380 m alt.

**ETYMOLOGY.** Named (noun in apposition) for Fosimondi, the only known site for the species. The survival of this species in in the hands of the Fosimondi people.

**SPECIMENS.** CAMEROON. Cameroon Highlands, Fosimondi, path leading from Fosimondi to Besali, fr. 28 Apr. 2005, *Tchiengué 2212* (K, YA); ibid., *Tchiengué 2213* (K, YA); Ntoo forest at Fosimondi, 5°38'03"N, 9°57'59"E, fl. 24 Feb. 2006, *Tchiengué 2583* (K, K-spirit, MO, P, YA); ibid., *Tchiengué 2584* (BR, K, WAG, YA); fl., fr. 24 Feb. 2006, *Tchiengué 2597* (K holo; BR, WAG, YA iso).

**CONSERVATION.** *Argocoffeopsis fosimondi* is here assessed as CR B2ab(iii), Critically Endangered according to the criteria of IUCN (2001) since it is only known from a single site (area of occupancy 4 km$^2$ using 4 km$^2$ cells) and since this site is under severe threat of forest clearance for cash and food crops (Tchiengué pers. obs. 2006). It is to be hoped that the people of Fosimondi, led by their elders and Fon, will seek to protect this beautiful shrub from extinction.

## ACKNOWLEDGEMENTS

We thank Jean-Michel Onana (IRAD-National Herbarium of Cameroon) for facilitating our research and the Darwin Initiative of the UK Government for part funding the fieldwork and research which led to this paper. The Overseas Fieldwork Committee of the Royal Botanic Gardens, Kew funded the fieldwork of April 2005 and February 2006. Mark Coode translated the

Latin. Diane Bridson reviewed an earlier version of this manuscript. The second author thanks the Fon and people of Fosimondi for hosting him and his plant surveys in 2005 and 2006. Aline Horwath, Bate Oben, Denis Ndeloh, Messrs Nwopiemba, Tanjong and Columbus helped find and collect specimens at Fosimondi. Louis Nkembe of ERuDeF encouraged us in this survey and supplied staff to assist.

## REFERENCES

Cheek, M., Pollard, B.J., Darbyshire, I., Onana, J.-M. & Wild, C. (eds) (2004). *The Plants of Kupe, Mwanenguba and the Bakossi Mountains, Cameroon: A Conservation Checklist.* Royal Botanic Gardens, Kew, UK. iv + 508 pp.

Davies, A.P. & Sonké, B. (2009). A new *Argocoffeopsis* (Rubiaceae from Southern Cameroon: *Argocoffeopsis spathulata. Blumea* 53: 527–532.

IUCN (2001). *IUCN Red List Categories and Criteria. Version 3.1.* IUCN, Gland, Switzerland. 30 pp.

FIG. 5 (opposite) *Argocoffeopsis fosimondi* **A** habit, flowering branch; **B** stem, showing transition from matt black juvenile to white, swollen, corky, older epidermis; **C** second internode from apex, showing stipule with gum exudate and supra-axillary buds; **D** inner surface of stipule showing gum deposits and absence of colleters; **E** 3-flowered inflorescence with corollas fallen; **F1** flower bud and attachment to stem (from spirit material); **F2** detail of **F1** showing calyculi; **G** flower bud, longitudinal section; **H** corolla and stamens, plan view (from spirit); **I** corolla and stems side view (one lobe removed); **J** corolla lobe flattened to show asymmetry (from spirit); **K** stamens, lateral (left) and ventral view (right); **L** fruit (immature); **M** fruit (mature); **N** fruit apex, plan view, showing absence of calyx limb, raised disc and style base; **O** seed, side view. **A, C & D** from *Tchiengué* 2584; **B, E, M & N** from *Tchiengué* 2597; **F–K & P** from *Tchiengué* 2583 (spirit); **O** from *Tchiengué* 2212. Scale bars: double, graduated = 5 cm; double ungraduated = 1 cm; single, graduated = 5 mm. Drawn by Andrew Brown.

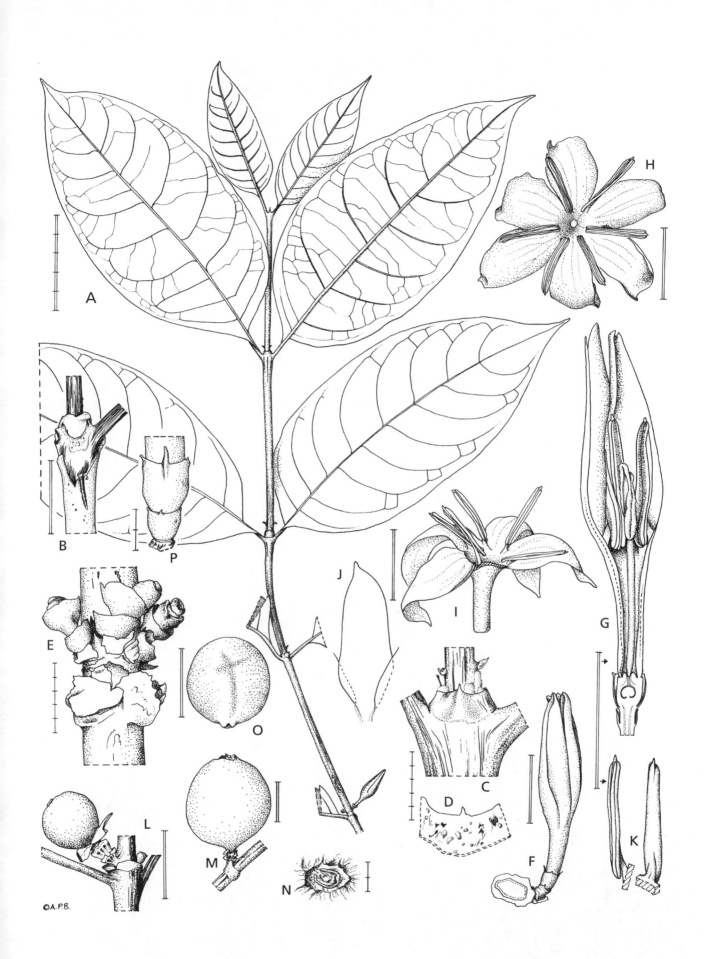

# *MYRIANTHUS FOSI* (CECROPIACEAE) A NEW SUBMONTANE FRUIT TREE FROM CAMEROON

Martin Cheek & Jo Osborne

Herbarium, Royal Botanic Gardens, Kew, Richmond, Surrey, TW9 3AE, UK

**Abstract:** *Myrianthus fosi* Cheek sp. nov. is validated as a submontane tree with edible fruits from the Cameroon Highlands in Cameroon and is assessed as Vulnerable using the 2001 IUCN criteria. Its taxonomic affinities are assessed.

De Ruiter (1976) revised *Myrianthus* P. de Beauvois and *Musanga* R.Br., the two African genera of the Cecropiaceae. Formerly they were included in the Moraceae. De Ruiter recognised seven species of *Myrianthus* which are all trees or shrubs (one is a liana) of evergreen forest. In the notes under *Myrianthus serratus* (Trécul) Benth. & Hook., de Ruiter (1976: 494) makes the following remarks about the taxon which is the subject of this article:

"Some collections recently made in Cameroun at altitudes between 850 and 1250 m (*Letouzey* 10 802, 11 208, 13 641, *Satabié* 105) are provisionally placed in *M. serratus*. These collections have sub-persistent stipules, like in *M. serratus* var. *serratus*. The cortex of the leafy twigs is red-brown like in *M. serratus* var. *letestui*. The leaves are mostly 3–5-lobed, whereas they are commonly entire in *M. serratus* var. *serratus*. The base of the lamina is acute or subcordate. The indumentum is brownish, while it is normally whitish in *M. serratus*. The petioles are up to 27 cm long and distinctly different in length on the same twig, like in *M. libericus*. The material is too scarce to allow a well-founded decision about its taxonomic position."

These remarks are repeated almost word-for-word, translated into French, in the Flore Du Cameroun account for Moraceae (including Cecropiaceae) by Berg *et al.* (1985: 270), with the addition of the specimen *Letouzey* 14532. In recent years, surveying plants of natural habitats in Cameroon, teams of botanists from RBG Kew and IRAD-National Herbarium of Cameroon have uncovered additional specimens with missing stages of this taxon, always in submontane forest. Most notable is *Osborne* 46, the first in fruit, revealing additional characters supporting the distinct specific status of this taxon. Accordingly it is here formally described and its affinities with other taxa are discussed. The description conventions of de Ruiter (1976) have been followed. All specimens cited have been seen unless otherwise indicated.

**Myrianthus fosi** Cheek **sp. nov.** a *M. serrato* (Trécul) Benth. & Hook. petiolis usque 27 cm longis in uno caule aut longis aut brevibus (non uniformibus), foliis palmatim 3–5(–7)-lobatis (non elobatis), infructescentiis 13–35 fructus gerentibus endocarpiis laevibus (non fructibus 1–4 endocarpiis longitudinaliter porcatis) differt. Typus: Cameroon, Littoral Region, Yabassi-Yingui area, proposed Ebo National Park, Research Station-West Transect, 675 m, fr. 12 Sept. 2006, *Osborne* 46 (holotypus YA, isotypi BR, K, P).

Syn. *Myrianthus* sp. 1 of Kupe, Cheek & Darbyshire p. 260 in Cheek *et al., The Plants of Kupe, Mwanenguba and the Bakossi Mountains, Cameroon: A Conservation Checklist* (2004).

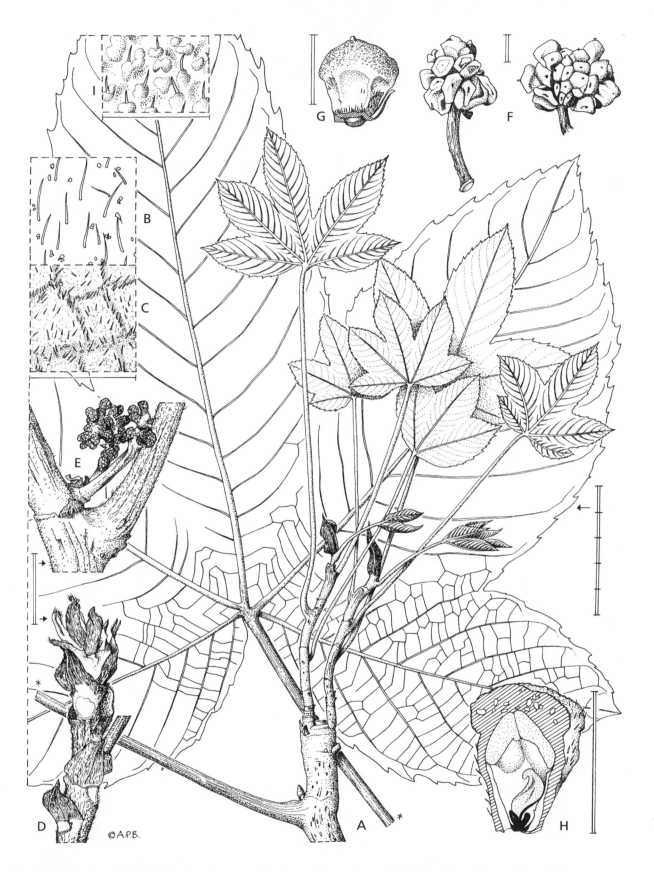

**FIG. 6** *Myrianthus fosi* **A** habit showing young leaves and one mature leaf (abaxial surface, with petiole cut at **); **B** indumentum of adaxial leaf surface (young leaf); **C** indumentum of abaxial surface (young leaf); **D** stem apex showing stipules; **E** male inflorescence; **F** infructescences; **G** fruit showing two attached bracts and below, exposed, base of the seed; **H** longitudinal section of a fruit; **I** fruit, upper surface with papillate hairs. **A-C & F-I** from *Osborne* 46; **D** from *Fenton* 280; **E** from *Letouzey* 11208. Scale-bars: single bar = 1 mm; double bar = 1 cm; graduated double bar = 5 cm. All drawn by Andrew Brown.

Tree 5–15 m tall, probably evergreen. Bark smooth, trunk 14.8 cm diameter at breast height (*Fenton* 280). Stems of leafy shoots scented of antiseptic or 'deep heat' when cut, terete, fistular, 8–9 (–12) mm diam., internodes 7–20 mm long when mature, outer surface matt red brown, with scattered raised concolorous elliptic lenticels 1.5–2 mm long, indumentum dense of appressed simple, soft, straight golden-brown hairs 0.1–0.5 mm long, soon glabrescent. Leaves alternate, simple, chartaceous to coriaceous, ± orbicular in outline, 15–30 cm long, 11–36 cm broad, base truncate to cordate, 3–5(–7)-lobed, lobing by $^1/_3$ –$^1/_2$(–$^2/_3$) of the radius of the leaf as measured between the outermost and its neighbouring lobe, lobes oblong-elliptic, apex broadly acute to rarely long acuminate, central lobe with 6–14 secondary nerves on each side of the midrib, nerves arching near margin towards leaflet apex, not terminating in a marginal tooth, lower surface of the blade brownish white, tertiary nerves scalariform, quaternary nerves reticulate, raised, subdivided by fifth and sixth order veinlets, nerves brown, interstitial areas brownish white, matt, with arachnoid hairs mixed with straight, transparent, patent 0.05–0.2 mm long hairs, upper surface subglossy, dark green, hairs sparse, straight, translucent (0.2–)0.3–0.4 mm long, margin serrate-dentate, teeth irregular, slightly hooked, apex mucronate, c. 1.5 teeth per cm, each 3–11 mm long; petiole 6–28 cm long, both long and short on the same stem at adjacent nodes, longitudinally ribbed, matt dark brown, indumentum as stem; stipules intrapetiolar, at length deciduous, subpersistent, clasping the stem by $^3/_4$ to $^9/_{10}$ of the circumference, ovate, 0.7–0.8 × 0.9–1.2 cm, midrib prominent, hairs appressed, moderately dense. Staminate inflorescences immature (including the peduncle) 1.15–1.5 × 0.4–0.8 cm with 2–4 primary branches, each one itself branching dichotomously; ultimate flowering units bearing c. 20 flowers (in bud only), ellipsoid, c. 2.5 × 2 mm; peduncle c. 1.3 cm long, 1–2 mm thick, indumentum of translucent patent simple hairs 0.1–0.3 mm long, the surface moderately densely covered with brown ellipsoid structures 0.02–0.03 mm long. Female inflorescence unknown. Infructescence with peduncle 30 × 5 mm, fruiting head 3–3.5 cm diam., with 13–22(–30) fruits; bracts oblong, c. 4 × 2.5 mm; fruits edible, ripening from green to yellow (*Osborne* 46), obovoid, 12 × 12 mm, proximal half 5–6-faceted due to mutual compression, appressed long-hairy, hairs 0.5–0.8 mm long, distal half convex or flat, mucron 1 mm, surface papillate-hairy, papillae (0.06–)0.11–0.18 mm diam., bearing a bristle hair (0.1–)0.15–0.18 mm long; endocarp ovoid-ellipsoid c. 8 mm, bony, pale yellow, smooth, lacking longitudinal ridges. Fig. 6

**DISTRIBUTION.** Cameroon, known only from Central Region, SW Region, W Region and Littoral Region.

**HABITAT.** Submontane forest with *Santiria trimera* (Burseraceae), *Allanblackia gabonensis* (Guttiferae), *Syzygium staudtii* (Myrtaceae); 832–1400 m alt.

**ETYMOLOGY.** Named (noun in apposition) in honour of Mrs Mary Fosi, former National Focal Point person for the Convention on Biological Diversity with the Ministry of the Environment and Forests of Cameroon who has long supported the project for a Red Data book for the Plants of Cameroon and who is herself, like this species, a native of the Cameroon Highlands.

**LOCAL NAMES AND USES.** Boussombossomb (Banen *fide* L. Beheng), fruit edible, also eaten by monkeys, chimps and gorillas (*Osborne* 46).

**SPECIMENS.** CAMEROON. Central Region: Ndikiniméki, 40 km S, entre Ndom et Ndambok, Massif de Nkokon, st. 13 Dec. 1971, *Letouzey* 10802 (K, P n.v., YA n.v); Ndikiniméki, 40 km NW,

Bankounga, ♂ infl. 12 Feb. 1972, *Letouzey* 11208 (K, P n.v., YA n.v.). SW Region: Mamfe, 55 km SW, Abakpa-Mbiofang, *Letouzey* 13641 (P n.v., YA n.v.); Rumpi Hills, 30 km NW Kumba, Lokando-Dikome Balue, st. 23 March 1976, *Letouzey* 14532 (K, P n.v., YA n.v.); Bakossi Mts, Kodmin, st. Jan. 1998, *Cheek* in *Plot Voucher Plot B* 234 (YA); Lebialem Highlands, Fosimondi at Ntoo forest 5°38'02"N, 9°57'44"E, st. 25 Feb. 2006, *Tchiengué* 2623 (K, YA). W Region: 12 km N Bafang, S/P Company, fl. buds, 12 Nov. 1974, 10°13'N, 5°13'E, 1250 m alt., *Satabié* 105 (P n.v., YA image viewed). Littoral Region, Yabassi-Yingui area, Ebo proposed National Park, Research Station-West transect, 675 m, fr. 12 Sept. 2006, *Osborne* 46 (holo K; iso BR, P, WAG, YA); ibid, 1360 m South transect LHS 0.66 m 4°20'45"N, 10°24'40"E, st. 9 Dec. 2007, *Fenton* 280 (BR, K, P, YA). Note that the record *Letouzey* 13641 has not been seen by the authors but is cited by de Ruiter (1976) and Berg *et al.* (1985).

**CONSERVATION.** *Myrianthus fosi* is here assessed as Vulnerable VU B2ab(iii) according to the criteria of IUCN (2001). Only eight sites are known equating to an AOO of 32 km$^2$ if 4 km$^2$ cells are used. The species is threatened at Lebialem Highlands since its forest habitat has been observed to have been cut down for farms over several years of study. While the sites in Rumpi Hills and Bakossi Mts are likely to be unthreatened for the moment due to low population pressure and remoteness from infrastructure, the stands of those near Bafang, Mamfe and Ndikiniméki is unknown and requires investigation. This appears to be a very sparsely distributed species even within its submontane forest habitat in the Cameroon Highlands. For example, although c. 9000 specimen numbers were made at Kupe-Bakossi, only the single specimen cited above was made of this species. However at a very local level it can be common since *Letouzey* 11208 remarked 'abundant' and *Letouzey* 14532 'frequent'. It may qualify as a rare pioneer since some collections derive from habitat which is disturbed, e.g. *Satabié* 105 from a young coffee plantation. Only at the Ebo site is more than one specimen known which may suggest that it occurs most densely here. Since there are prospects of formal protected status at Ebo, and a long-term conservation project there it seems logical that further research and conservation efforts should be focussed here. However efforts should be make to educate conservation managers and local communities at other sites to recognise and protect *Myrianthus fosi*.

## TAXONOMIC AFFINITIES

De Ruiter (1976: 494) compared our taxon with *M. serratus* var. *serratus* and *M. serratus* var. *letestui* de Ruiter, pointing out several differences and similarities which we have expanding in Table 1 below with the benefit of fruiting material which shows additional taxonomic characters unavailable to de Ruiter.

*Myrianthus cuneifolius* (Engl.) Engl. considered closely related to *M. serratus* by de Ruiter, is also included in this table. Both species occur in lower Guinea. De Ruiter also mentioned a similarity with a disjunct species in West Africa, *M. libericus* Rendle. The two taxa both have petioles which are distinctly different in length on the same stem, and are very similar in many respects of their leaf shape and size. Perhaps for this reason Letouzey named most of his specimens of *M. fosi* as *M.* cf. *libericus*. However the two can be distinguished using the features in Table 2 below, and so de Ruiter seems justified in rejecting Letouzey's hypothesis. By the time Letouzey collected his last

**TABLE 1.** The more significant characters separating *Myrianthus fosi* from *M. serratus* and *M. cuneifolius*.

| | *Myrianthus serratus* | *Myrianthus fosi* | *Myrianthus cuneifolius* |
|---|---|---|---|
| Leaves | simple, entire | simple 3–5(–7)-lobed | compound, 3–7 leaflets or simple, entire |
| Indumentum of stems | hairs straight, appressed, translucent 0.01–0.1 mm long, smooth to touch | hairs straight appressed to patent, golden, 0.1–0.5 mm long, smooth to touch, not scabrid | hairs erect, curved, translucent, 0.25 mm long, scabrid |
| Petiole length variation on a stem | ± constant | both short and long | ± constant |
| Number of fruits per infructescence | 1–3(–7) | 13–22(–30) | 3–6 |
| Fruit apex | acute | convex or flat-topped | acute to rostrate |
| Endocarp surface | longitudinally ridged | smooth | smooth |

**TABLE 2.** The more significant characters separating *Myrianthus libericus* from *Myrianthus fosi*.

| | *Myrianthus libericus* | *Myrianthus fosi* |
|---|---|---|
| Stem epidermis | grey | matt red brown |
| Leaves | simple, entire or 3(rarely 5)-lobed | simple, 3–5(–7)-lobed |
| Stipules | early caducous | subpersistent |
| Male inflorescences | (3–)5–11 cm long | 1.5 cm long |
| Infructescence fruit number | 3–6 | 13–22(–30) |
| Fruit shape | ovoid | obovoid |
| Geography | Guinea (Conakry) to Ghana | Cameroon |
| Habitat | Lowland forest | Submontane forest |

specimen of this taxon (*Letouzey* 14532, 23 March 1976), perhaps by then aware of de Ruiter (1976) he determined it as *Myrianthus* sp. and observed "without doubt a species identical to the trilobed form of other submontane areas". Clearly he was aware that this was a distinct taxon lacking a name by that stage.

*Myrianthus fosi* is unlikely to be confused with any other species in the field since it is confined to submontane altitudes in Cameroon where no other *Myrianthus* occurs apart from *Myrianthus preussii* Engl. which always has compound leaves. The latter taxon has an unusually wide altitudinal range, occuring from 500–1400 m alt. being most frequent in the submontane part of this band.

Several collections attributed to *M. serratus* in Cameroon, all sterile, resemble *M. fosi* but occur in lowland swamp forest (*Letouzey* 12636, 14931). Further investigation might reveal that these are conspecific with *M. fosi*. Similar habitat disjunctions occur in other species such as *Symphonia globulifera* L. and *Pseudagrostistachys africana* (Muell. Arg) Pax. & K. Hoffm. which also occur in both lowland swamp forest and submontane forest, but not generally in lowland forest.

It is possible that this species has a future as a fruit crop since its fruits are reported to be edible (*Osborne* 46) and since the related *M. arboreus* P. Beauv. has fruits known as the 'bush pineapple' which are widely enjoyed in Cameroon.

Further work especially additional fertile specimens, are still needed since *M. fosi* remains an incompletely known taxon. The female inflorescence and male inforescences at anthesis are still unknown to science. The origin of the aromatic scent recorded by *Fenton* 280 and *Osborne* 46 is worth investigating. No previous observation of scent appears to have been made in *Myrianthus* (de Ruiter 1976).

Variation in leaf lobing is notable. In the Rumpi Hills specimen, Letouzey (14532) noted that he had only seen the 5-lobed variant here and that elsewhere in submontane areas (see his other specimens cited above) leaves were 3-lobed. However both forms can occur in one forest since at Ebo, *Fenton* 280 is 3-lobed and *Osborne* 46 is 5-lobed. Leaf lobing variation can be correlated with age of the plant, or with the switch to sexual reproduction. However, while *Letouzey* 11208 (male inflorescence) is 3-lobed, *Osborne* 46 (fruiting) is 5-lobed. Entire, unlobed leaves are very rare. The authors have only observed one among all the specimens examined, that being on a duplicate of *Fenton* 280. Otherwise all the other leaves of this number seen are 3-lobed. This may be simply a juvenile leaf since de Ruiter (1976: 473) states that seedlings of *Myrianthus* start with simple leaves which gradually pass into the leaves characteristic for the adult specimens.

## ACKNOWLEDGEMENTS

Mark Coode is thanked for the Latin translation, Jean-Michel Onana of IRAD-National Herbarium of Cameroon for facilitating fieldwork permits. Bethan Morgan for supporting fieldwork in the Ebo Forest. Abwe Enang and Leopold Beheng for assisting the type collection and Zachary Bekokon for providing information about the fruits. Two anonymous botanists are thanked for reviewing the manuscript. The first author's contribution was made under the Darwin Initiative supported 'Red Data Plants Cameroon' project.

## REFERENCES

Berg, C. C., Hijman, M.E.E. & Weerdenburg, J.C.A.(1985). Moracées. *Flore Du Cameroun* 28. MESRES, Yaoundé.

Cheek, M., Pollard, B.J, Darbyshire, I., Onana, J.-M. & Wild, C. (eds) (2004). *The Plants of Kupe, Mwanenguba and the Bakossi Mts, Cameroon: A Conservation Checklist.* Royal Botanic Gardens, Kew, UK. iv + 508 pp.

de Ruiter, G. (1976). Revision for the genera *Myrianthus* and *Musanga* (Moraceae). *Bull. Jard. Bot. Nat. Belg.* 46: 471–510.

IUCN (2001). *IUCN Red List Categories and Criteria. Version 3.1.* IUCN, Gland, Switzerland. 30 pp.

# RED DATA TAXA OF LEBIALEM HIGHLANDS

Martin Cheek

Herbarium, Royal Botanic Gardens, Kew, Richmond, Surrey, TW9 3AE, UK

## INTRODUCTION

As in the previous conservation checklists in this series (Harvey *et al.* 2004, Cheek *et al.* 2004, Cheek *et al.* 2010), **all** flowering plant taxa recorded in the checklist have been assessed on a family by family basis for their level of threat, i.e. as threatened (CR — critically endangered; EN — endangered; VU — vulnerable), near threatened (NT) or of least concern (LC).

The main part of this chapter consists of taxon treatments, giving detailed information on the 42 Red Data species known to be present in the checklist area. Several of these taxa were assessed for the first time in the process of writing this book while others had been assessed as threatened in other publications and are reassessed here. IUCN rules do not allow acceptance of taxa unless they are either published or on the brink of publication. Consequently, the several new species to science that are not yet at this point, some of which are only known from Lebialem Highlands, have not been assessed and are therefore not mentioned in this chapter.

This, the introductory part of the chapter, details the methodology used in making the assessments. The Red Data taxa are detailed by vegetation type in the chapter on Vegetation.

## ASSESSMENTS — METHODOLOGY

### TAXA OF LEAST CONCERN (LC)

In the first place all taxa that were found to be fairly widespread e.g. Extent of Occurrence (IUCN 2001) greater than 20,000 km² (and/or from 20 or more localities) were listed as LC. These facts were established principally using FWTA as an indication of range and number of collections sites, supplemented by other published sources, such as Flore du Cameroun, or by research into specimens at K, in cases of doubt. Taxa which, by these measures, are widespread and common do not qualify as threatened under Criterion B of IUCN (2001), the main criterion used in our previous checklists for assessing threatened species. Under Criterion A, widespread and common taxa may, in contrast, still be assessed as threatened, but only if their habitat, or some other indicator of their population size has been, and/or is projected to be, reduced by at least 30%, in the space of three generations, so long as this does not exceed 100 years.

### THREATENED AND NEAR THREATENED TAXA

Those taxa in the checklist that were not assessed as LC, were then checked for level of threat using IUCN (2001). Most of these taxa were at least fairly narrowly endemic (restricted) in their distributions, e.g. endemic to Cameroon, or to NW Region, Cameroon and Nigeria. Criterion C, which demands knowledge of the number of individuals of a species, was not used, since these data were not usually available. Criterion E, which depends on quantitative analysis to calculate the probability of extinction over time, was also not used. Criterion D was used for two species which although rare, appear to face no immediate threats: *Begonia schaeferi* and *Plectranthus punctatus* subsp. *lanatus*.

## i) Using Criterion B

About five-eighths of the assessments (27/42) were made using Criterion B (usually B2ab(iii)), since the nature of the data available to us for our taxa lends itself to this criterion. Knowledge of the populations and distributions of most tropical plant species is mainly dependent on the existence of herbarium specimens. This is because there are so many taxa, most of which are poorly known and have never been illustrated. For this reason observations based only on sight-records are particularly unreliable and so undesirable in plant surveys of diverse tropical forest. Exceptions can be made when a family or genus specialist is present, working with a monograph at hand, or a proficient tree spotter identifying timber tree species. In contrast, surveys of birds and primates are not specimen dependent, since species diversity in these groups is comparatively low, and comprehensive, well-illustrated identification guides are available.

For the purpose of Criterion B we have almost always taken herbarium specimens to represent 'locations'. Deciding whether two specimens from one general area represent one or two locations is open to interpretation, unless they are from the same individual plant. Generally in the case of several specimens labelled as being from one town e.g. 'Bipinde', or one forest reserve e.g. 'Bafut Ngemba FR', these have been interpreted as one 'location'. Where a protected area has been divided into several geographical subunits, as at Mt Cameroon (see Cable & Cheek, 1998), and it is known that, say six specimens of a taxon occurring at Mt Cameroon fall into two such subunits, then this is treated as two locations. 'Area of occupancy' (AOO) and 'extent of occurrence' (EOO) have been measured by extrapolating from the number of locations at which a species is known. The grid cell size used for calculating area of occupancy is that currently advocated by IUCN, that is 4 km$^2$. Information on declines in Criterion B has been obtained from personal observations, sometimes supplemented by local observers.

## ii) Using Criterion A

The remaining assessments, about three-eighths (15/42) of the total, were made using Criterion A. This criterion was used because montane and submontane taxa of the Cameroon Highlands that extend to the Lebialem/Bamboutos/Bamenda Highlands are relatively easy to assess thanks to RBG Kew GIS unit studies of forest loss in two parts of the Bamenda Highlands, the first with c. 25% loss over 8 years in the Kilum-Ijim area (Moat in Cheek *et al.* 2000) and the second, with c. 50% loss over 15 years, in the Dom area (Baena in Cheek *et al.* 2010).

These data have been used to estimate that during the last century, over 30% of the forest habitat of montane (above 2000 m alt.) and submontane (800–2000 m alt.) species has been lost in the Cameroon Highlands as a whole, so qualifying species with this range as Vulnerable. *Xylopia africana* is an example of such a species. Formerly such taxa were often treated as unthreatened (e.g. Cable & Cheek 1998) because they are secure on Mt Cameroon where little montane forest loss has occurred. Losses in this montane forest habitat have been highest in the Bamenda

Highlands (and Bamboutos Mts) where c. 96.5% of original forest cover has been lost (Cheek in Cheek *et al.* 2000). Therefore species restricted to this range have been estimated as having lost up to 80% of their habitat over the last century.

### iii) Using Criterion D

Assessments using Criterion D generally depend on a knowledge of the global numbers of mature individuals (D). This knowledge is generally not available for plant taxa in Cameroon except under VU D2, which can be applied to taxa which, although not obviously immediately threatened, are nevertheless rare in terms of total numbers of individuals or area of occupancy.

## CHANGES IN IUCN CRITERIA

It is to be hoped that there will be a moratorium on the changes in IUCN guidelines which have been made in recent years. These changes made the work of assessors more difficult and have also reduced comparability of assessments made in different years. Assessments made in earlier Cameroonian checklists (e.g. Cheek *et al.* 2000) under IUCN (1994) criteria have been updated according to IUCN (2001) criteria where the taxa occur in the present checklist.

## RED DATA TAXON TREATMENTS

The following Red Data taxon treatments are mostly modified and updated from accounts written earlier, both published (e.g. The Plants of Bali Ngemba, Harvey *et al.* 2004) and unpublished (e.g. the Red Data book of Cameroon Plants).

# DICOTYLEDONAE

**ACANTHACEAE** (I. Darbyshire)

*Brachystephanus giganteus* Champl.
(syn.: *Oreacanthus mannii* Benth.)
VU B2b(iii)c(iv)
**Range:** Bioko (2 coll.); Cameroon (W Cameroon: Kupe-Bakossi (2 coll.); Rumpi Hills (1 coll.); Mt Cameroon (13 coll.); Lebialem-Bamboutos (1 coll.); Bamenda Highlands (Mt Oku (3 coll.) & Bafut Ngemba (1 coll.)).
A species of the gregarious, mass-flowering herbaceous community of upland forest, it is more restricted in altitudinal range than *Acanthopale decempedalis*. It appears locally common on Mt Cameroon but is scarce elsewhere, being restricted to Mwanenguba and on Mt Kupe in the current checklist area. Both are pre–1980 collections, and this species was not recorded during extensive inventory work at these sites in the 1990s, perhaps because these surveys did not coincide with mass-flowering years. The assessment of this taxon made in Cheek *et al.* (2004: 139), based on its habitat loss exceeding 30% in c. 30 years, is maintained here since no new data are available apart from the Lebialem record that does not change the assessment.
Previously named *Oreacanthus mannii* Benth., this taxon has been reassigned to *Brachystephanus* by D. Champluvier in her recent monograph of this genus.

**Habitat:** closed canopy montane forest; (1300–)2000–2600 m alt.

**Threats:** much of the montane forest in SW Region remains relatively undisturbed, but clearance for agriculture is widespread in the Bamenda Highlands, threatening these populations. Furthermore, the cyclical mass-flowering habit of this species results in large fluctuations in mature populations, making it susceptible to short-term stochastic change, for example local fire events or landslides which could adversely affect seedling populations.

**Management suggestions:** current population sizes in terms of both number of individuals and area of occupancy should be assessed as this is not clear from the specimen data. A better understanding of the flowering cycle of this community should be gained, including population fluctuations between cycles. Bamenda Highlands populations outside protected areas may provide valuable information on how immature populations respond to increased anthropogenic pressures.

As for *Acanthopale decempedalis*. In addition, a survey of this taxon should be carried out during the next mass-flowering year on Mt Kupe and Mwanenguba.

## ANNONACEAE (M. Cheek)

### *Xylopia africana* (Benth.) Oliv.
VU A2c

**Range:** São Tomé (1 coll.); SE Nigeria (Obudu (1 coll.)); Cameroon (Mt Cameroon (4 coll.), Bakossi (10 coll.), Fosimondi (1 coll), Bali Ngemba (numerous coll.)).

Of all the forests that we have surveyed, *Xylopia africana* has been found in greatest density at the Bali Ngemba Forest Reserve in North West Region. This, now the largest remnant of the forest that cloaked the Bamenda Highlands at the 1300–1900 m range, is only 100 Ha in extent and shrinking fast due to illegal clearance for farming.

Presumably this species was once common in the Bamenda Highlands where it is now all but extinct. While there are no figures for rates of forest loss in the Bamenda Highlands as a whole, in one area which has been studied, the Kilum-Ijim area, forest loss of 25% over 8 years in the 1980s–1990s has been recorded (Moat in Cheek *et al.* 2000). Past forest loss in the Bamenda Highlands is therefore the main basis for the threat to *Xylopia africana*. On Mt Cameroon it appears rare, being found only twice in the surveys of 1992–1994. Elsewhere in the mountains of the Cameroon line it is also known from the extension into Nigeria: the Obudu Plateau where it is also threatened due to forest clearance, if indeed, it is still extant there. It is also known from São Tomé in the Gulf of Guinea. Strangely, it is not known from Bioko, the Rumpi Hills or the Bamboutos Mts. Mt Kupe and the Bakossi Mts probably now support the largest single subpopulation of *Xylopia africana*.

**Habitat:** submontane and lower montane forest; 800–2000 m alt.

**Threats:** clearance of forest for timber and agricultural land.

**Management suggestions:** if proposals to protect forest above 1000 m alt. in much of Bakossi are enacted and respected, this subpopulation seems secure.

## ARALIACEAE (M.Cheek)

### *Schefflera hierniana* Harms
VU B2ab(iii)

**Range:** Bioko; Cameroon (Belo to Lake Oku, Rumpi Hills (1 coll. each); Mt Cameroon, Bakossi Mts, Lebialem Highlands at Fosimondi (2 coll. each)).

A strangling epiphytic shrub of cloud forest resurrected by David Frodin from synonymy with the much commoner *S. barteri*, this rare species is known from only six sites along the wetter forested parts of the Cameroon Highland chain. It was assessed as vulnerable in Cheek *et al.* 2004: 145 and that assessment is maintained here although Tchiengué has since found it in the Fosimondi area, where forest is extremely threatened. In future it is likely that this species will be rated as endangered since it seems to be genuinely rare within its range. It is separated from *S. barteri* by having a slender 1–2 cm acumen, persistent fleshy-leathery bracts at the base of the inflorescence which are most easily seen in immature inflorescences, and congested densely brown-scurfy pubescent inflorescence axes, partial-peduncles 2–3 mm long, pedicels 2 mm (Frodin & Cheek in Cheek *et al.* 2004: 246)

**Habitat:** evergreen forest; 900–1400(–2100 m) alt.

**Threats:** forest clearance for agriculture and wood. It has not recently been seen at the Belo to Lake Oku site despite intensive surveys in the late 1990s (Cheek *et al.* 2000) and since forest has almost disappeared between these two locations, it is probably extinct there. Habitat degradation has been steady on some parts of Mt Cameroon at these altitudes and the species was not refound there in intensive surveys of the early 1990s. At Fosimondi forest is disappearing very rapidly to create farms.

**Management suggestions:** the best hope for the survival of this species may be the Bakossi Mts at the Kodmin and Nzee Mbeng sites since pressure here is relatively low and since the species was seen here recently (1998). Continued surveys of the unexplored parts of the Bakossi Mts might well reveal new sites. Further surveys at its other sites, if focussed on this species, might yet reveal that it survives so long as significant areas of forest exist.

## BALSAMINACEAE (M. Cheek)

### *Impatiens letouzeyi* Grey-Wilson

EN B2ab(iii)

**Range:** Cameroon (Bakossi Mts (5 coll.), Lebialem Highlands-Fosimondi (1 site)).

This robust, epiphytic herb has the largest flowers of all Cameroonian, and perhaps African, *Impatiens*. Described in 1981 from a single specimen in western Bakossi collected in the 1970s (*Letouzey* 15353), it was rediscovered by an Earthwatch team led by George Gosline in 1998 and subsequently found to be fairly common in the Kodmin-Edib area. At the end of the wet season, plants are readily spotted by their fallen flowers under the trees on which they grow. This plant has not been found on Mt Kupe despite searches in all months over several years. It is arguable as to whether the Bakossi records should be taken as two or four sites. Opting for the second gives a total number of three sites (AOO 12 km² using 4 km² cells, threats as below). The assessment above is maintained from that given in Cheek *et al.* (2004). The additional record from Lebialem both extends the range and increases the threat level to the species but does not take the species out of the EN status.

**Habitat:** epiphytic in crown of trees 4–6 cm from ground, in shrubs over streams or very rarely, terrestrial, on lakeside *Sphagnum* blanket; 1200–1350 m alt.

**Threats:** the planned reservoir scheme near Kodmin may threaten part of the population of this species. At Fosimondi the trees on which this species is growing are being cut down to plant crops.

**Management suggestions:** a detailed study of this species at Kodmin-Edib where the species is most secure and where conservation efforts might be focussed, would provide more precise data on population density and demography, as well as assisting in the placement of future development schemes. Botanical surveys in other parts of the Bakossi Mts, or Rumpi Hills, might discover more sites which would reduce the threat assessment of this taxon.

*Impatiens sakeriana* Hook.f.

VU A2c

**Range:** Bioko; Cameroon (Mt Cameroon (6 coll.), Mwanenguba (2 coll.), Fosimondi-Bamboutos Mts (1 coll.), Bamenda Highlands (1 coll.), Bali Ngemba Forest Reserve [FR] (3 coll.), to Mt Oku (12 coll.)).

This often robust, presumed perennial, locally common terrestrial herb has the highest altitudinal range of all Cameroonian *Impatiens*. Secure on Mt Cameroon, and probably Bioko. However, forest at Mwanenguba, the Bamenda Highlands and Bamboutos Mts has been under pressure from grassland fires set by graziers and for clearance for agriculture; forest in the Kilum-Ijim area outside the protected area having seen a reduction in cover of c. 25% in eight years of the 1980s and 1990s (Moat, rear cover in Cheek *et al.* 2000). Assuming a generation time of ten years, it is estimated that about 30% habitat loss may have occurred in the last 30 years. This species is uniformly present in montane forest above c. 2000 m alt., occurring at a higher elevation than any other species of the genus in the Cameroon Highlands. The assessment made in Cheek *et al.* 2004 is maintained here.

**Habitat:** understorey of montane forest; 2000 m alt.

**Threats:** see above.

**Management suggestions:** enforcement of protected area boundaries. Demographic studies are needed to elucidate generation time and ecological requirements of this taxon.


## BEGONIACEAE (M.Cheek)

*Begonia adpressa* Sosef (Sect. *Loasibegonia*)

VU B2ab(iii)

**Range:** Cameroon (Kupe-Bakossi, Rumpi Hills, Mt Nlonako, S Bamboutos, Lebialem Highlands (8 pre–1995 coll.)).

Described by Sosef in 1992, when only eight collections were known (Sosef 1994: 150), all of which were from W of Bangem in in the northern Bakossi Mts apart from three collections immediately to the North in the Bamiléké area, one to the west, in the Rumpi Hills, and one immediately to the east, at Mt Nlonako. To these we added six further collections from three sites made since 1995, one from Kupe village, one from Nyasoso, the others all from Kodmin (Cheek *et al.* 2004). This constitutes nine locations and an area of occupancy of 14 km$^2$ calculated at 1 km$^2$ per collection. The habitat quality at the Bakossi sites is projected to remain stable if proposed conservation methods come to pass, but all other sites except the Rumpi Hills natural forest at this taxon's altitude range are under severe pressure, or have been eliminated. The conservation assessment of this species in Cheek *et al.* 2004 is maintained here since no additional data are available apart from an extra record at Lebialem Highlands where its habitat was found to be very highly threatened (Harvey *et al.* 2010).

**Habitat:** terrestrial and on rocks in forest; 1000–1750 m alt.

**Threats:** forest clearance for timber, followed by agriculture, especially at Nlonako, Bamboutos and Lebialem sites where active and ongoing.

**Management suggestions:** the future for *Begonia adpressa* seems bleak outside of the Rumpi Hills and Bakossi Mts. However, at these last sites it seems secure if proposed conservation measures are respected.

***Begonia oxyanthera*** Warb. (Sect. *Tetraphila*)

VU A2c

**Range:** Nigeria (1 coll.); Bioko (6 coll. from 2 sites); Cameroon (Mt Cameroon (4 coll.), Mt Kupe and the Bakossi Mts (5 coll.), Rumpi Hills (1 coll.), Mwanenguba (1 coll.), Lebialem Highlands/ Bamboutos Mts (2 coll.), Bamenda Highlands (17 coll.)).

The range data above are taken from the account of the species in the recently published revision of *Begonia* sect. *Tetraphila* (de Wilde 2002). In previous Red Data assessments (Cable & Cheek 1998 and Cheek *et al.* 2000) this taxon was listed as LR nt. Here it is reassessed as vulnerable because of habitat destruction in what appears to be its main subpopulation in the Bamenda Highlands from which most of the c. 30 specimens listed by de Wilde derive. Moat (in Cheek *et al.* 2000) records forest loss in the Kilum-Ijim protected area itself. Since the 'generation' duration of this taxon might easily be five years, it is estimated that habitat loss for the species over its area of occupancy as a whole is likely to have been over 30% in the last 15 years. The conservation assessment of this species in Cheek *et al.* 2004 is maintained here since no additional data are available apart from its discovery at Lebialem, where its habitat is very severely threatened.

The increasing density of this species, moving northwards (Mt Cameroon and Bioko only c. 2 sites each; Bamenda Highlands numerous sites and collections) is perhaps a reflection of a longer dry season requirement.

**Habitat:** submontane and montane forest; 1200–2200(–2400)m alt.

**Threats:** forest clearance for wood and agriculture (mainly in the Lebialem/Bamboutos/Bamenda Highlands).

**Management suggestions:** subpopulations in SW Region, and probably also in Bioko, are fairly secure, lacking threats. In NW Region the substantial subpopulation in the well protected Kilum-Ijim site (Mt Oku and the Ijim Ridge) may be the only locality where the taxon will survive, unfortunately, thus protection here is important.

***Begonia preussii*** Warb. (Sect. *Tetraphila*)

VU A3c

**Range:** Nigeria (3 coll.); Bioko (1 coll.); Cameroon (Edea-Kumba-Lebialem (15 coll.)).

This taxon was previously assessed as endangered (Cable & Cheek 1998) but is here downrated to vulnerable in the light of the taxonomic revision by de Wilde (2002, source of the range data above) showing it to be more common and widespread in Cameroon than was previously thought. In addition, the discovery of eight further specimens at five sites in Mt Kupe-Bakossi Mts confirms the foregoing. Nonetheless, due to the low altitudinal range of this taxon, habitat loss of about 30% over the next 15 years (estimated as equating to three generations) can be postulated. The conservation assessment of this species in Cheek *et al.* 2004 is maintained here since no additional data are available apart from a new record at Lebialem which needs further confirmation since it extends the range of the species North considerably, and has a much higher alt. (1486 m) than all others. Its forest habitat at this site is extremely threatened.

**Habitat:** evergreen forest; 850–1000 m alt.

**Threats:** clearance of forest for wood and agricultural land is a major threat throughout its range and probably accounts for the lack of collections from Bioko in the last century (where forest was largely cleared below 1000 m alt.).

**Management suggestions:** this taxon is most likely to survive in lowland forest reserves such as the Bakossi FR and Mokoko FR, but only if their boundaries are respected. Resources need to be found to achieve this.

***Begonia pseudoviola*** Gilg (Sect. *Loasibegonia*)
VU B2ab(iii)
**Range:** Cameroon (Lebialem/S Bamboutos Mts (6 coll.), Kupe-Bakossi (3 coll.), Mt Nlonako (3 coll.)).

Described by Gilg in 1904 from *Conrau* 10 collected in Nov. 1898 between Banti and Babesong at 600–700 m alt., this beautiful species is easily recognised by its diminutive size and its often purple-black leaf-blades which bear long, white, patent-curved hairs on their upper surfaces.

The nine collections recorded by Sosef (1994: 182) are immediately to the north-east (Nkongsamba-Mt Nlonako) or North (S and W Bamboutos Mts) of Kupe-Bakossi. Until we worked at Ngomboaku in 1999, recording three sites for the species in a fairly small area east of the village, *B. pseudoviola* was entirely unknown from Kupe-Bakossi. Owing to its relatively low altitudinal range, *B. pseudoviola* is extremely vulnerable to forest clearance (see discussion at beginning of this chapter). For the purposes of this assessment we rate this species as VU under criterion B, being known from ten locations and having an area of occupancy of 12 km$^2$ (see *B. adpressa*). The prognosis for habitat destruction for this taxon is high. This taxon might be better assessed under criterion A, but lack of data on the state of sites west of the Bamboutos Mts made this difficult to apply until recently. In 2009 observations at Lebialem proved the species to be extremely threatened west of the Bamboutos. However specimens from that location are slightly anomalous, having green not black leaves and having a lower range (250–400 m) so need further scrutiny. The conservation assessment of this species in Cheek *et al.* 2004 is maintained here since no significant additional data are available.
**Habitat:** wet mid-elevation forest; (250–)600– 850 m alt.
**Threats:** see above.
**Management suggestions:** revisting the Ngomboaku site to assess the population size and potential threats should be made; active protection should be sought at this location.

***Begonia schaeferi*** Engl. (Sect. *Loasibegonia*)
VU D2
**Range:** Nigeria (Obudu Plateau (1 coll.)); Cameroon (Mwanenguba, Kongoa Mts, Mt Nlonako (each 1 coll.); Lebialem Highlands/Bamboutos Mts (3 coll.); Bamenda Highlands (3 coll.)).

First collected in Nov. 1900 at Bare, Mwanenguba, by Schaefer and described in 1921, the species has not since been seen in the Kupe-Bakossi area and may be extinct there. Personal observations of *B. schaefferi* in the Bamenda Highlands shows that this species is demanding in its habitat requirements and likely to occur at extremely few sites within a given area (Cheek *et al.* 2000).

The assessments of this species as VU D2 in Cheek *et al.* (2000, 2004) are maintained here: no new data are available on the taxon apart from a recent record from Lebialem/Bamboutos where its habit is extremely threatened (Harvey *et al.* 2010). The information presented below is mainly taken from those works.
**Habitat:** on rocks and vertical rock faces in moist to comparatively dry places in primary submontane to montane forest, the latter sometimes with trees not taller than 6–12 m; c. 1500–2300 m alt. (Sosef 1994).

**Threats:** while cliff faces generally are unlikely to be disturbed, clearance of adjoining forest for fuel and agriculture could endanger this species by removing the shade necessary for its survival.

**Management suggestions:** during surveys of cliff spaces (in forest) this species should be looked for and, if located, the number of plants and locality recorded.

## CECROPIACEAE (M.Cheek)

### *Myrianthus fosi* Cheek
VU B2ab(iii)

**Range**: Cameroon (Central: 40 km S Ndikiniméki, ibid, Massif de Nkokon; SW: Mamfe, 55 km SW, Abakpa-Mbiofang, Rumpi Hills, Bakossi Mts, Kodmin, Lebialem Highlands, Fosimondi; W: 12 km N Bafang; Littoral: Yabassi-Yingui area, Ebo proposed National Park).

*Myrianthus fosi* is here assessed as Vulnerable according to the criteria of IUCN (2001). Only eight sites are known equating to an AOO of 32 km$^2$ if 4 km$^2$ cells are used. The species is threatened at Lebialem Highlands since its forest habitat has been observed to have been cut down for farms over several years of study. While the sites in Rumpi Hills and Bakossi Mts are likely to be unthreatened for the moment due to low population pressure and remoteness from infrastructure, the stands of those near Bafang, Mamfe and Ndikiniméki is unknown and requires investigation. This appears to be a very sparsely distributed species even within its submontane forest habitat in the Cameroon Highlands. For example, although c. 9000 specimen numbers were made at Kupe-Bakossi, only the single specimen cited above was made of this species. However at a very local level it can be common since *Letouzey* 11208 remarked 'abundant' and *Letouzey* 14532 'frequent'. Only at the Ebo site is more than one specimen known which may suggest that it occurs most densely here.

**Habitat:** submontane forest with *Santiria trimera* (Burseraceae), *Allanblackia gabonensis* (Guttiferae), *Syzygium staudtii* (Myrtaceae); 832–1400 m alt.

**Threats:** see above.

**Management suggestions:** since there are prospect of formal protected status at Ebo, and a long-term conservation project it seems logical that further research and conservation efforts should be focussed there. However efforts should be make to educate conservation managers and local communities at other sites to recognise and protect *Myrianthus fosi*.

## CELASTRACEAE (I. Darbyshire, B.J. Pollard & M. Cheek)

### *Salacia lebrunii* Wilczek (assessed by M. Cheek)
VU B2ab(iii)

**Range**: Cameroon (SW: Bechati-Lebialem); Gabon (33 km E of Lastourville; Belinga iron mine); Congo (Brazzaville) (Chaillu); Congo (Kinshasa) (Kinshasa, Lac Leopold II; Lodja-Kole; Yangambi). Known from six localities (AOO 24 km$^2$ with 4 km$^2$ cells) and with threats as below, *Salacia lebrunii* is here assessed as Vulnerable.

This liana has 4-ridged stems and elliptic leaves c. 7–9 × 3.2–5 cm, including a 0.5 cm acumen, lateral nerves are 3–6 pairs. The flowers are single, petals light orange, 2 × 1 mm, stamens only 2 (usually 3 in *Salacia*); fruit subglobose, hard, orange, 2–3 cm diam.

**Habitat**: lowland forest; 100–400 m alt.

**Threats**: Belinga is set to become a Chinese-owned iron mine. At Lebialem-Bechati, forest clearance for agriculture in ongoing.

**Management suggestions**: in Chaillu threats are few and Ogooué-Leketi proposed NP (the site of this species) may be the best focal point for conserving the species. The status at the other sites is unknown. The elders at Bechati should be informed of the presence and conservation importance of this species.

*Salacia lehmbachii* Loes. var. *pes-ranulae* N. Hallé (assessed by I. Darbyshire & B. J. Pollard)
VU B2ab(iii)

**Range:** Nigeria (Cross River State: Oban (1 coll.)); Cameroon (SW: Bakossi FR (1 coll.), Lebialem (1 coll.), 15 km S of Akwaya (1 coll.); S: S of Ebolowa (3 coll.)).

First described in 1986, all previous collections of this taxon were made prior to 1980, when lowland forest South of Ebolowa appeared to be the centre of its distribution. These forests have experienced significant reductions following expansion of lowland plantation agriculture. The discovery of this taxon in the protected Bakossi FR is of significance to its future conservation. This assessment was originally made in Cheek *et al.* 2004 and is maintained here since only one extra record (Lebialem) has been added since. Six sites are known (AOO 24 km$^2$ with 4 km$^2$ cells). *Salacia lehmbachii sensu lato* is distributed from Sierra Leone to Tanzania; the complex comprises seven varieties of which var. *pes-ranulae* is both the most localised in distribution and the most distinct morphologically.

**Habitat:** dense lowland forest understorey; 450–700 m alt.

**Threats:** continued loss of lowland forest throughout the species range, particularly in Cross River State, Nigeria and the forests of S and SW Regions, Cameroon. Illegal encroachment of agriculture into the Bakossi FR, facilitated by European Community road building, threatens this population.

**Management suggestions:** heightened protection of remaining forest at the Bakossi FR may help to preserve this taxon at Bakossi. Further botanical inventory work in lowland sites, such as the neighbouring Loum FR, may reveal further populations; care should be taken to accurately identify all future collections of *S. lehmbachii* to variety level.


## CHRYSOBALANACEAE (M. Cheek)

*Magnistipula conrauana* Engl.
EN A3c

**Range:** Cameroon (Lebialem-Bamboutos Mts (5 coll.), Mwanenguba (1 coll.), Kupe-Bakossi (4 coll. at 4 sites)).

Letouzey and White (Fl. Cameroun 20, 1978) record the collections for this taxon as restricted to the Bamboutos-Mwanenguba area. During the late 1990s we found the taxon also to occur, rarely although fairly widespread, at Mt Kupe and the Bakossi Mts. It is estimated here that this canopy tree will become extinct in the wild at Bamboutos and Mwanenguba (given the threats indicated below) in the next 100 years but has a good possibility of surviving in the newly discovered southern half of its range.

The assessment here, made in Cheek *et al.* 2004, is maintained since no new data are available apart from the record at Lebialem. A Missouri record from Korup (*Gentry* 52789) is regarded as very unlikely, being far outside the altitudinal range of the species. Two records from the Rumpi Hills by Thomas, are more likely, but these are based on field determinations only and need verification before they can be accepted.

**Habitat:** submontane forest, 1000–1500 m alt.

**Threats:** the Bamboutos Mts are densely populated and intensively cultivated, for such crops as *Coffea arabica*. What few fragments of forest remain are under great pressure to supply wood and for further agricultural land. The rate of forest area loss of 25% in eight years for the Oku area cited elsewhere is probably exceeded in the Bamboutos area, and this rate of loss appears to be continuing unabated.

**Management suggestions:** Letouzey & White (1978 *loc. cit.*) mention this plant occurring in hedges in farmland in Bamboutos. Encouragement of this practice, and also evaluating sacred forests there for the existence of this species, might be the means to survival in the northern part of the range. In the southern part of its range *M. conrauana* seems likely to survive if existing conservation plans go ahead for the Bakossi NP.


## EUPHORBIACEAE (M. Cheek)

***Pseudagrostistachys africana*** (Müll.Arg.) Pax & K.Hoffm. subsp. ***africana***
VU A2c

**Range:** Ghana; São Tomé; Bioko; SE Nigeria (Obudu Plateau (1 site)); Cameroon (Mt Etinde, Mt Kupe and Bakossi Mts, Bali Ngemba FR, Mefou proposed NP).

Listed as VU A1c, B1 + 2c by Hawthorne in 1997 (www.iucnredlist.org ( IUCN 2003)), this monotypic genus, a tree restricted to submontane forest (apart from at one lowland site in Ghana and one S of Yaoundé at Mefou proposed NP), probably now has its largest subpopulation in Bakossi, where it is fairly secure. It was reassessed as VU A2c, B2ab(iii) in Cheek *et al.* 2004 on the basis of more extensive disturbance data in Cameroon and according to the modified IUCN criteria of 2001. That assessment is maintained as far as criterion A, but the B assessment is dropped since more than 10 sites are now known now that it has been discovered at Mefou and Lebialem. It is estimated that more than 30% of the population has been lost in the last 100 years due to submontane or swamp forest habitat destruction in major parts of its range, principally Ghana, Nigeria and, in Cameroon, the Bamboutos and Bamenda Highlands. This destruction is ongoing. This is one of several species showing habitat disjunction between lowland swamp and submontane forest. In the field it is readily recognisable by the large Irvingiaceae-like sheathing apical stipule and the long-scalariform tertiary leaf venation.

**Habitat:** submontane, or rarely lowland swamp forest; 500–1500 m alt.

**Threats:** forest clearance for wood and agriculture (Obudu Plateau, Bamboutos/Bamenda Highlands: Lebialem and Bali Ngemba FR).

**Management suggestions:** the status of this taxon in São Tomé and Bioko needs more investigation. Bali Ngemba FR represents the most easily accessible and dense population of the taxon, followed by Mt Kupe and Mt Etinde; these are the more promising sites for demographic studies of the taxon that would inform management planning. It is possible that surveys in lowland swamp forest will find more sites for this species.


## GUTTIFERAE (M. Cheek)

***Allanblackia gabonensis*** (Pellegr.) Bamps
VU A2c

**Range:** Cameroon (Kupe-Bakossi (11 coll.); Bali Ngemba, Bamenda Highlands (several coll.);

Lebialem-Bamboutos: Fosimondi, Mt Bana, Batcham; Central: Yaoundé, Ebolowa, Sangmelima (1 coll. each)); Gabon (Moubighou, Moucongo and Tcyengue (1 coll. each)).

This tree is conspicuous for carpeting the forest floor with its pale lemon-coloured fallen flowers (usually red in the Lebialem and Bali Ngemba subpopulations), each about 6 cm across. The largest part of its domain was probably once the Bamenda Highlands, where submontane forest is now confined to a few small parcels, the largest of which is at Bali Ngemba. Forest in these highlands is still being lost: 25% of forest in one area disappeared in an eight-year period ending in 1995 (Moat in Cheek *et al.* 2000). Overall, more than 30% of the habitat of this species has probably disappeared over the last 100 years. Mt Kupe and the Bakossi Mts are now probably the stronghold for *A. gabonensis*. Elsewhere the species occurs on several of the small hills dotted through the forest belt in South and Central Region, finally extending into the Crystal Mts of Gabon.

**Habitat:** submontane forest; 700–1500 m alt.

**Threats:** continued forest clearance for agriculture and wood particularly in the Lebialem/ Bamboutos/Bamenda Highlands and Yaoundé area.

**Management suggestions:** this species is fairly secure in the upper part of its altitudinal range in Kupe-Bakossi, but enforcement of the forest reserve boundary is needed if it is to survive at Bali Ngemba. Surveys should be conducted on the forested hills from which it has been collected elsewhere in Cameroon and also in the Crystal Mts, to determine whether it survives there and whether it can be protected at any of these sites.

## *Psorospermum aurantiacum* Engl.

VU B2ab(iii)

**Range:** Nigeria (Obudu Plateau (5 coll.) and Mambilla); Cameroon (SW: Lebialem; NW: Bambui, Bamenda, Kumbo-Oku, Bafut-Ngemba, Bali Ngemba (12 collections); W: Kounden (1 coll.)).

This tree or shrub is distinctive for the dense orange-brown hairs on the lower surface of the leaf, the upper surface a contrasting glossy black when dried. It appears confined to the Bamenda Highlands, with outliers in the adjoining Obudu Plateau and Lebialem/Bamboutos Mts (Kounden). The assessment here was made in Harvey *et al.* 2004 and is maintained here. Six sites can be accepted if the Bamboutos/Bamenda Highlands are taken as one large subpopulation. Clearly a 4 km$^2$ cell size is inadequate there. Overall an AOO of 80 km$^2$ is estimated. In future this species could be better assessed under criterion A2c since its distribution and habitat corresponds with that which has seen most loss in Cameroon.

**Habitat:** edge of gallery forest; 1500–1800 m alt.

**Threats:** dry-season grassland fires, usually set by man, burn into the montane and submontane forest in the Cameroon Highlands, reducing its area. It is possible, even likely, that *P. aurantiacum* by the nature of its habitat, has some resistance to fire and may even benefit from occasional fires. However, the current frequent and intense fires may affect individuals adversely. Conversion of forest to farmland, by contrast, is an undoubted threat. Over 25% of forest in one sample area of the Bamenda Highlands was lost in the 1980s–1990s (Moat in Cheek *et al.* 2000).

**Management suggestions:** research to explore the effect of different fire regimes on this species is advised. In the short term, the highest priority is to re-find individuals at the known sites and to seek means to protect these. While almost all natural forest at Bafut-Ngemba FR has disappeared already, Bali Ngemba FR still remains fairly intact and may be the best prospect for the conservation of *P. aurantiacum*.

## ICACINACEAE (M. Cheek)

*Pyrenacantha longirostrata* Villiers
EN B2ab(iii)

**Range**: Cameroon (SW: Mt Cameroon-Mokoko, Kumba-Limbe, Lebialem-Bechati, Akwaya-Mamfe); Gabon (Monts de Cristal).

Known from the five sites above, each from a single collection, *Pyrenacantha longirostrata* is here assessed as Endangered (AOO 20 km$^2$ with 4 km$^2$ cells and threats below).

Easily recognised by its fruits with a beak 1–2 cm long. This is a glabrous-stemmed, 6 m liana with elliptic leaves 13–18 × 4.3–8 cm. The long acumen bears a mucro. Secondary nerves are 5–6 pairs. 6–8 flowers occur at the apex of the 6–9 cm spike-like lateral inflorescence. The 3 mm-long flowers are 4-merous.

**Habitat**: lowland evergreen forest c. 200–300 m alt.

**Threats**: highly threatened at most known lowland sites in Cameroon by logging followed by agriculture, especially at Kumba-Limbe where it may already have been lost by forest clearance, and at Mokoko Forest Reserve (scheduled for logging in 2010) and Bechati (pers. obs.). The threat status of the species at Monts de Cristal and Akwaya-Mamfe is unknown.

**Management suggestions**: there is no obvious site that might be the focus for conservation of *P. longirostrata*. One option is to seek to refind plants at the most recently discovered site, at Bechati (Sept. 2006). Here the elders of Bechati should be consulted to secure the protection of the species, and data gathered on regeneration levels, density and range of the species.


## LABIATAE (M. Cheek)

*Plectranthus punctatus* L'Hér. subsp. *lanatus* J. K. Morton
VU D2

**Range:** Cameroon (Lebialem/Bamboutos Mts (5 coll.) and Bamenda Highlands (four. pre-1996 specimens)).

This subspecies was first collected by Maitland, probably either on Laikom Ridge or at 'Mbesa Swamp' (*Maitland* 1724, "Basenako-Lakom, on plateau in grassland, June 1931, 1800 m"). In the Bamenda Highlands it is also known from Ndu and from the Bambili Lakes as well as Kilum-Ijim.

This plant is best distinguished from other members of the genus by the purple-blotched, white hairy stems and small stature. Its habitat is also distinctive.

Herb c. 30 cm tall. Stems ascending, rounded and sub-succulent when alive, nodes swollen, internodes 1.5–3(–5) cm long, whitish green, blotched purple, clothed in long white hairs c. 2 mm long. Older stem bases sometimes straggling, prostrate, rooting at nodes. Leaves of the midstem sessile, sublanceolate, c. 6 × 2 cm, apex rounded, margin serrate-crenate, with c. 15 teeth per side, white hairy above and below. Inflorescence 6–15 cm long, verticils 5–10, internodes c. 1.5 cm long, flowers subsessile. Flower with corolla c. 12 mm long, pale blue, prominently speckled with purple.

The largest population of this species seen was at Afua swamp at Kilum-Ijim in November 1999 where it lines the border of the swamp for about 100 m or more. In total, five sites for the species are known at Kilum-Ijim, the numbers of individuals of which varies from 5–10 to 100–200.

This assessment was originally made in Cheek *et al.* 2000 and is maintained here since no data changes are available apart from an extra record from Lebialem.

**Habitat:** damp grassland at the edge of swamps, or banks; 1800–2600 m alt.

**Threats:** swamp drainage or development; shading-out by growth of grasses; possibly trampling by cattle.

**Management suggestions:** more research is needed on the management regime needed for this plant. However plants seen in long grass at the Mbesa swamp in 1998 seemed less healthy and were far fewer than those seen in close cropped grass at Afua swamp.

## LECYTHIDACEAE (I. Darbyshire & M. Cheek)

*Napoleonaea egertonii* Baker f.

VU B2ab(iii)

**Range:** Nigeria (Cross River State, Oban (3 coll.)); Cameroon (SW: Atolo to Mamfe (1 coll.), Kupe-Bakossi (3 coll., 3 sites), Nta Ali (1 coll.), Takamanda Forest Reserve (1 coll.), Korup NP (1 coll.), Lebialem-Bechati (1 coll.)); Gabon (Mahounda (1 coll.), Ikembélé (1 coll.)).

This striking medium-sized (c. 40 cm diam. breast height) forest tree , bearing large (to 70 cm × 30 cm) spiny fruits on the trunk, was only known from Oban Nigeria, prior to the plant inventory work in western Cameroon beginning in the 1980s. Discoveries of this species at Korup is important as it is relatively well protected at that site; however, it is not common at Korup, only 1–2 trees having been found (Cheek, *pers. obs.*), and its abundance at Takamanda is unknown. At Kupe Village and the adjacent Manehas Forest Reserve, the species is again uncommon, one plant being found at each location. However, several specimens were observed within close proximity to Nyandong in W Bakossi. The assessment made here first featured in Cheek *et al.* 2004 and is maintained here since only the Lebialem site has been added since. The assessment is supported on the basis of 12 spots (AOO 48 km² with 4 km² cells) at nine locations and threats as below.

**Habitat:** low- to mid-elevation evergreen forest, often occurring on rocky slopes; 250–1000 m alt.

**Threats:** the Nigerian sites are likely to have been either lost or under severe threat from widespread logging of lowland forest here. This is the case at all other known sites, however threats in Gabon are unknown. The two sites at Mt Kupe are below the 1000 m alt. lower limit of effective forest protection and thus vulnerable to agricultural encroachment. At Nyandong, several trees were recorded close to the village and adjacent to tracks; these are highly vulnerable to future expansion of the village and road improvement.

**Management suggestions:** a survey of the number of trees of this species at Korup should be carried out to determine its abundance at this site, as it is the best protected, thus offering the best opportunity for conservation of this species. Informing local communities, most notably at Nyandong where this species appears most common and in closest proximity to human settlement, of the scarcity of this species may help to promote community-led conservation, particularly as it is such a striking and easily recognisable taxon. Enquiries are needed to determine whether or not the large and numerous seeds have any local uses which might encourage local protection.

## LOGANIACEAE (M. Cheek)

*Anthocleista scandens* Hook.f.

VU A2c

**Range:** Bioko (Clarence Peak, Moca); SE Nigeria (Obudu (1 site)); São Tomé (5 sites); Cameroon (Gepka, Nkambe (1 coll.); Mt Oku (2 sites); Bafut-Ngemba (2 coll.); Lebialem-Fosimondi, Mt Kupe (9 coll. at 3 sites); Mt Etinde (3 coll.)).

It is estimated here that, over the last 100 years, over 30% of the submontane forest habitat of this species, mostly in the Bamenda Highlands and Bamboutos, has been lost due to forest clearance (see threats below).

In flower this strangling epiphytic shrub or small tree is spectacular with its clusters of large bright white flowers.

**Habitat:** submontane forest; 1200–2000 m alt.

**Threats:** forest clearance for agriculture and wood, particularly in the Bamenda Highlands, where forest loss has been running at 25% over eight years at one sample area (Moat in Cheek *et al.* 2000). forest losses in recent decades suggest that *A. scandens* may no longer survive at Bafut-Ngemba, Gepka or Obudu. Forest at Bafut-Ngemba for example has been completely replaced by *Eucalyptus* (pers. obs. Cheek 2004) and clearance is ongoing at Fosimondi (Tchiengué pers. comm. 2005–6).

**Management suggestions:** conservation efforts are probably best focussed where sites are most concentrated, namely São Tomé (5 sites, but protection levels unknown) and Mt Kupe (3 sites, protection levels currently high). At Mt Oku this taxon occurs at the lower altitudinal boundary of protection and is likely to become extinct; one of the two sites there occuring outside this boundary. The population at Mt Etinde occurs at the peak and is inaccessible to all but trekkers. Unless the summit area is cleared for touristic purposes, *A. scandens* seems secure here.

## MELASTOMATACEAE (M. Cheek)

*Cincinnobotrys letouzeyi* Jacq.-Fél.

EN B2ab(iii)

**Range:** Cameroon endemic (SW: Mamfe, 45 km ENE at Numba; Bechati in Lebialem Highlands). From a creeping rhizome like a string of beads, this species produces an erect single elliptic leaf blade c. 12 × 5 cm on an 8–12 cm long petiole and an inflorescence almost as long of c. 12 pink flowers c. 2 cm wide.

Here *C. letouzeyi* is assessed as endangered since only two sites (AOO 8 km$^2$ with 4 km$^2$ cells) with threats as below.

**Habitat:** on rock near spray from streams in evergreen forest belt.

**Threats:** forest destruction at the Bechati side.

**Management suggestions:** protection at either of the two sites is desirable. Data in population size, density and regeneration levels is desirable.

*Dissotis bamendae* Brenan & Keay

VU B2ab(iii)

**Range:** Nigeria (Mambilla Plateau, Nguroje FR (1 coll.)); Cameroon (NW: Bamenda (2 coll.), Nchan (1 coll.), Bafut-Ngemba FR – Bali Ngemba FR (3 coll.); W: Bamboutos-Bamenda Highlands (Santa, 1 coll.); SW: Lebialem-Fosimondi).

This many-stemmed, shapely shrub was sunk into *D. princeps*, a widespread species (South Africa to Ethiopia) by Jacques-Félix, but since it is abundantly distinct (e.g. 3 leaves per node, not 2) we follow FWTA in maintaining it as different. Only 9 sites are known (AOO 36 km$^2$ with 4 km$^2$ cells, and threats as below). The assessment made by Cheek in Harvey *et al.* 2004 is maintained here, since only one extra record (Lebialem) is added.

**Habitat:** grassland, often near streams, possibly requiring proximity to forest; 1500– 2200 m alt.

**Threats:** poorly known, but including conversion of natural habitat to farmland, whether for grazing or for cultivation of beans or potatoes. If plants are dependent upon proximity to forest, then ongoing loss of surviving forest in the Bamenda Highlands is a major threat.

**Management suggestions:** long term studies of exclusion plots to study the effect of removing fire and/or grazing from this species would help to understand management regimes for this and other ecologically similar species in the Bamenda Highlands, but would need a site where the species is frequent.

## MELIACEAE (M. Cheek)

*Heckeldora ledermannii* (Harms) J. J. de Wilde

EN B2ab(iii)

**Range:** Cameroon endemic (SW: Mts Ekomane (Kongoa), Fosimondi and Bakossi Mts).

A shrub or small tree up to 5 m high, 5–9-foliolate leaves, glabrous apart from the puberulous midrib, calyx <1 mm long, young inflorescences 12–30 cm long, fruits epiculate but not beaked, nor ribbed, nor moniliform.

Here *H. ledermannii* is assessed as endangered since three sites are known (AOO 12 km$^2$ with 4 km$^2$ cells) with threats as below. The type locality, Kongoa Mts, have not been visited by botanists in many decades.

**Habitat:** submontane evergreen forest; 900–1500 m alt.

**Threats:** forest clearance for agriculture and wood at lower altitudes only on Mt Kupe, but wholesale at the Fosimondi site where extremely threatened by farm expansion.

**Management suggestions:** *Heckeldora ledermannii* appears secure in Bakossi, especially at Mt Kupe where it occurs at high density (given the number of specimens made there), so long as submontane forest continues to be protected here, the species is likely to survive.

## RUBIACEAE (M. Cheek)

*Argocoffeopsis fosimondi* Tchiengué & Cheek

CR B2ab(iii)

**Range:** Cameroon endemic (SW: Cameroon Highlands, known only from Fosimondi-Lebialem Highlands near W Bamboutos Mts).

Assessed in Harvey *et al.* 2010 as Critically Endangered according to the criteria of IUCN (2001) since it is only known from a single site (area of occupancy 4 km$^2$ using 4 km$^2$ cells) and since this site is under severe threat of forest clearance for cash and food crops (Tchiengué pers. obs. 2006).

**Habitat:** submontane (cloud) forest with *Santiria trimera* (Burseraceae), *Cola verticillata* (Sterculiaceae), *Macaranga occidentalis* (Euphorbiaceae), *Chassalia laikomensis*, *Pauridiantha paucinervis* (both Rubiaceae), *Carapa procera* (Meliaceae); 1330–1380 m alt.

**Threats:** see above.

**Management suggestions:** it is to be hoped that the people of Fosimondi, led by their elders and Fon, will seek to protect this beautiful shrub from extinction. Educational material on the importance of this material will be passed to them by the authors via ERuDeF.

*Chassalia laikomensis* Cheek

CR A2c

**Range:** Nigeria (Mambilla Plateau (1 coll.)); Cameroon (Mwanenguba (1 coll.), Lebialem, Bamenda Highlands (several sites: Bali Ngemba, Ijim and Dom)).

The assessment above was made in Cheek & Csiba (2000), listed as having been assessed by Cheek *et al.* (2000) in IUCN (2003) (www.iucnredlist.org). That assessment was maintained in Cheek in Harvey *et al.* 2004 and is maintained here also. Lebialem is added here to its range. Several records from Bakossi of a similar plant may represent a separate taxon.

This is a forest understorey shrub, probably long-lived.

**Habitat:** montane evergreen forest; 1650–2000 (–2400)m alt.

**Threats:** about 95% of the original forest cover of the Bamenda Highlands has been lost to e.g. agriculture (Cheek *et al.* 2000; Cheek & Csiba 2000) and there have been similar losses at Mambilla and Mwanenguba. In the Dom area 50% of forest cover was lost in the 15 years 1988–2003 (Baena in Cheek *et. al.* 2010).

**Management suggestions:** more information is needed on the numbers of individuals at the known sites and levels of regeneration. Enforcement of existing protected area boundaries would help protect a significant portion of the surviving population.


*Coffea montekupensis* Stoff.

VU B1ab(iii) + B2ab(iii)

**Range:** Cameroon endemic (SW: Mts Kupe & Bakossi, and Lebialem Highlands – Fosimondi)

Assessed by Davis *et al* 2006: 492 as VU B1ab(iii). We accept and maintain this assessment here even though in Cheek *et al* 2004: 374 we assessed the species at NT.

*Coffea montekupensis* is readily recognised by its pink flowers. African *Coffea* species usually have white flowers. Within its extent of occurrence it is often rather frequent, reflected in the 39 specimens cited in Cheek *et al* 2004: 374, and the seven communities within the forests of which it was found. We have not measured its EOO accurately but believe it to be less than the 20,000 km$^2$ required to trigger B1. In 2005/06 Tchiengué discovered this species as Fosimondi — a significant range extension to the North. Therefore it is likely that it once extended between Fosimondi and Bakossi and may yet be found in this area where suitable habitat survives.

**Habitat:** submontane evergreen forest; 700–1500 m alt.

**Threats:** in Bakossi it is threatened by agriculture extending uphill from the fertile valley bottoms. At Fosimondi heavy forest losses have been observed by Tchiengué in successive years because of clearance for agriculture.

**Management suggestions:** the newly gazetted Bakossi National Park probably contains this species in a similar density to elsewhere in Bakossi and is likely to provide a long term future for the species. Meanwhile forest on the Mt Kupe, where it is best studied and documented, appear secure above 800–1000 m altitude and local communities should be supported to maintain this protection.


*Pavetta brachycalyx* Hiern

EN B2ab(iii)

**Range:** Cameroon (Mt Cameroon (9 coll. from 2 sites), Kupe-Bakossi (3 coll. from 2 sites), Lebialem-Fosimondi (1 site)).

Once thought restricted to Mt Cameroon, the range of this nondescript species was extended to Bakossi (Cheek *et al.* 2004: 177 where this assessment, maintained here, was made), in recent years a further site, at Fosimondi has been found. Five sites (AOO 20 km$^2$ with 4 km$^2$ cells) and threats as below justify EN status.

**Habitat:** lowland and submontane forest; 300–1500 m alt.

**Threats:** clearance of forest for agriculture especially in the lower part of the altitudinal range. Its recent discovery at Fosimondi reveals that its range extends to areas where forest survival is highly threatened.

**Management suggestions:** this species is reasonably secure where it occurs above c. 1000 m alt. in both the Bakossi and Etinde parts of its range. Elsewhere it is threatened. Populations outside of protected areas could be assisted in their survival by a poster campaign.

## *Pavetta hookeriana* Hiern var. *hookeriana*

VU A2bc

**Range:** Cameroon (Mt Cameroon (numerous coll.), Mwanenguba (c. 3 coll.), Bamboutos/Bamenda Highlands (numerous coll. at several sites: Lebialem, Bali Ngemba, Kilum-Ijim and Dom)).

It is estimated that over 30% of the habitat of this woody forest understorey species has been lost in the last century.

This is the highest altitude *Pavetta* known W of the Congo basin. Generally a good indicator of montane forest at 2000 m, its occurrence of altitudes of 1500–1700 m at Lebialem, Bali Ngemba and Dom is puzzling.

The assessment above is updated from that in Cheek *et al.* (2004) and Cheek in Harvey *et al.* (2004).

**Habitat:** montane forest; (1500–)1900–2000(–2400) m alt.

**Threats:** secure from threat at Mt Cameroon, *P. hookeriana* is threatened by forest clearance for agriculture and wood throughout the extensive Bamenda Highlands, probably once the main area for the species. Study of one large area in the highlands between 1987–1995 showed that 25% of the surviving forest was lost (Moat in Cheek *et al.* 2000) and elsewhere 50% was lost 1988–2003 (Baena in Cheek *et al.* 2010).

**Management suggestions:** improved policing of existing forest reserve boundaries could prevent extinction of this species in the Bamenda Highlands, where its survival is precarious, except at Kilum-Ijum. At Mwanenguba and Bamboutos Mts (presence inferred) it may not survive for much longer. The species is most secure at Mt Cameroon, where the narrowly endemic variety *pubinervia* also occurs.

## *Psychotria babatwoensis* Cheek
## (syn.: *Psychotria* sp. A of Bali Ngemba checklist)

EN B2a,b(iii)

**Range:** Cameroon (SW: Lebialem Highlands; NW: Bali Ngemba FR, Baba 2 community forest).

*Psychotria babatwoensis* is restricted to two, almost adjacent, small patches of submontane forest in the Bamenda Highlands of Cameroon and one in the Lebialem Highlands of SW Region about 30 km away. The first three collections were made in the Bali Ngemba Forest Reserve, SW of Bali. This forest is about 10 km$^2$ in extent and the understorey, the habitat of this species, is being steadily cleared for the planting of crops such as cocoyams (*Colocasia esculenta*). At this location the species was fairly rare (Cheek pers. obs.). Four other collections were made from the very much smaller community forest of the village of Baba 2, perhaps 2 km$^2$ in extent, probably the stronghold of the species, where it is relatively common. At Lebialem it is highly threatened by clearance for farms below the village of Fosimondi. The species has not been found in any of the few other surviving submontane forest fragments in the Bamenda Highlands, such as those of Dom or Mbiame, but is to be hoped that in future it will be located elsewhere. Since the area of occupancy of this species at these two sites can be no more than 12 km$^2$ and there is habitat loss due to past, ongoing, and expected future clearance at Bali Ngemba FR, *P. babatwoensis* was assessed when first published (Cheek *et al.* op. cit. 2009) as Endangered. Since no new data are available this assessment is maintained here.

**Habitat:** "Submontane forest with *Pterygota mildbraedii*" (Cheek in Harvey *et al.* 2004: 17); 1310–1750 m alt. *Psychotria babatwoensis* is one of the species characterising the above vegetation type, accompanying species being *Alangium chinense, Kigelia africana, Carapa grandiflora, Turreanthus africanus, Strombosia scheffleri, Euclinia longiflora* and *Chlorophytum sparsiflorum*.

**Threats:** see above.

**Management suggestions:** conservation efforts should be concentrated with the traditional authorities of the communities at the only known locations for this species, Lebialem Highlands at Fosimondi in SW Region, and Baba 2 and those around Bali Ngemba FR since these hold the future of the species in their hands. These authorities should be provided with the information needed to identify and protect this species. Data on regeneration levels and population size is needed to develop a plan to manage the species.

## *Sabicea xanthotricha* Wernham

VU B2ab(iii)

**Range:** SE Nigeria (Oban); Cameroon (SW: Mt Cameroon at Etinde and Mokoko, Bakossi Mts (Kodmin), Lebialem-Bechati; Littoral: Ebo).

Previously *S. xanthotricha* was assessed as EN B2ab(iii) since only four sites were known (Cheek *et al.* 2004: 181–182). New sites have been discovered at Ebo and Fosimondi, raising the total to six sites, hence AOO 24 km$^2$ at 4 km$^2$ cells. Accordingly we downrate the threat assessment to VU. In future it is likely to revert to EN or even CR since so many of its sites are threatened with habitat destruction and none are formally protected.

A pithy shrub in a genus of lianas.

**Habitat:** lowland and submontane evergreen forest; 0–1400 m alt.

**Threats:** forest clearance for agriculture and wood, particularly at Oban and at Mokoko FR.

**Management suggestions:** a survey to rediscover plants of this taxon at the known sites and gather the usual data on each of the subpopulations is recommended. The site at Ebo might offer the best prospect for conservation being placed in a proposed National Park.

## *Trichostachys petiolata* Hiern

(syn.: *Trichostachys* **sp. 1 of Kupe**)

EN B2ab(iii)

**Range:** Cameroon endemic (Littoral: 'Cameroons River' = Wouri R. (type); SW: Bakossi-Nyale, Bechati-Lebialem)

First collected in Jan. 1861 by the Kew botanist Gustav Mann. *T. petiolata* was not identified again until 2010, as this book was being assembled, when *van der Burgt* 885 from forest near Bechati, was identified as this species.

The species is unusual in its 7–8 cm long, non-interupted spike with obovate, c. 10 × 5 cm leaves, the apices rounded, petioles 1 cm long. Assessed here as Endangered since only three sites are known (AOO 12 km$^2$ using 4 km$^2$ cells) with threats as given below.

**Habitat:** lowland evergreen forest; 400–650 m alt.

**Threats:** forest clearance followed by agriculture e.g. at Bechati (van der Burgt pers. comm.)

**Management instructions:** none of the sites occurs in a formally protected area. The exact location of the type collection is unknown, while that in Bakossi, at Nyale is outside the new NP. Conservation efforts should be focussed at the only two known sites where neighbouring communities should be consulted and advised oin the rarity of this species and their help sought in protecting it.

## SAPINDACEAE (M. Cheek)

### *Allophylus bullatus* Radlk.

VU A2b,c

**Range:** Príncipe & São Tomé; SE Nigeria; Cameroon (Mt Cameroon, Mt Kupe, Lebialem, Bali Ngemba, Mt Oku and Ijim Ridge, Dom).

This understorey tree of upper submontane to montane forest, while secure on Mt Cameroon and on Mt Kupe, has lost large tracts of its habitat in recent decades in the Bamenda Highlands. Over 30% of its overall habitat is estimated to have been lost in the last 100 years. This taxon has also been assessed in Cheek *et al.* (2004), Cheek in Harvey *et al.* (2004) and Cheek *et al.* (2010). Most collections of this species are from Mt Cameroon and from Kilum-Ijim (Mt Oku and the Ijim Ridge) where it appears commonest. *Allophylus bullatus* is much rarer at lower altitudes where the bullate characteristic is often absent. Only 1–2 specimens are known each at Bali Ngemba and Dom. This species is distinctive in having white hairy domatia in the axils of tertiary nerves on the secondary nerves.

**Habitat:** upper submontane and montane forest; 1600–2400 m alt.

**Threats:** clearance of forest for agriculture and wood, particularly in the Bamenda Highlands of Cameroon, once probably the main habitat for *A. bullatus*. Study of one area here, the surroundings of Mt Oku (Moat in Cheek *et al.* 2000) showed that 25% of forest was lost between 1987–1995 and elsewhere, at Kejojang (Kejodsam), 50% was lost 1988–2003 (Baena in Cheek *et al.* 2010).

**Management suggestions:** improved policing of the existing protected areas might secure the future of this species. Support from local communities in protecting this species and its habitat should be sought and assisted.

### *Allophylus conraui* Gilg
(syn.: ***Allophyllus* sp. 1 of Kupe Bakossi checklist**)

EN B2a,b(iii)

**Range:** Cameroon (SW: Bakossi Mts, Banyang Mbo, Mone FR, "Kebo", Lebialem-Bechati).

The assessment made here is maintained from that in Cheek M. & Etuge, M. (2009): *Allophylus conraui* (Sapindaceae) reassessed and *Allophylus ujori* described from western Cameroon. Kew Bull. 64(3): 495–502. Since that publication only one other location, at Mone FR has come to light. Here it is locally common at several spots.

An area of occupancy of 20 km² is estimated for the species, based on a grid cell size of 4 km² and five sites being known. Threats are given below. Accordingly *A. conraui* is here maintained as Endangered, EN B2a,b(iii).

False records: Harris (2002: 189) records *Allophylus conraui* from the Central African Republic, but indicates some doubt as to the delimitation of the taxon. Subsequent observations (Harris pers. comm. to Cheek 2008) of the specimens on which the record is based, *Harris* 3308 and 4815 (E) indicates that these represent another taxon. The CAR taxon has hairs only 0.3–0.4 mm (not 3 mm) long, inflorescences twice as long as broad (not c. four times as long as broad); infructescences accrescent, 11.5 cm long (not non-accrescent, 2–4 cm long). Jongkind (2006: 379) cites *A. conraui* as occurring in Gabon, based on *Reitsma* 2996 and *Wilks* 2721. Observations of the first (the second specimen was not available), loaned from WAG to K, shows it to be another taxon, differing in the hairs <0.5 mm long (not 3 mm long); leaflets papery, marginal teeth inconspicuous, 1–2 per side (not membranous, marginal).

**Habitat:** lowland evergreen forest; 200–1000 m alt.

**Threats:** the site in the lowlands of W Bakossi Mts between Banyemem and Ayon is vulnerable to logging as is that to the N, in the Banyang Mbo Reserve which has had its protected status downrated. At Bechati in the Lebialem Highlands, the plant is under threat of forest clearance for agriculture, while there is a possibility that Mone FR will be legally logged.

**Management suggestions:** any moves to log the Mone FR, the site of the largest and most important subpopulation of *A. conraui* should be rejected since this is likely to destroy the habitat of this rare species. This site is the logical choice for the focus of conservation efforts for this species.

*Deinbollia oreophila* Cheek
(syn.: *Deinbollia* **sp. 1 of Kupe & Bali Ngemba**)
VU A2c
**Range:** Nigeria (Obudu plateau); Cameroon (Cameroon Highlands from Mt Cameroon to Bali Ngemba FR).

The assessment above, which appeared with the original publication of this species is maintained here (Cheek, M. & Etuge, M. (2009): A new submontane species of *Deinbollia* (Sapindaceae) from Western Cameroon and adjoining Nigeria. Kew Bull. 64(3): 503–508). *Deinbollia oreophila* was rated as VU A2c, i.e. vulnerable, given the continued loss of its submontane forest habitat, rated at 30% over its whole range over the last 100 years. In the northern part of its range, recent loss of its surviving forest habitat varies from 25% in 8 years to 50% in five years.

**Habitat:** understorey of submontane evergreen forest; (880–)1000–2050 m.

**Threats:** forest clearance for agriculture, especially at Mt Cameroon-Bambuko, Obudu, Lebialem-Fosimondi, Bali Ngemba, where it is active and ongoing.

**Management suggestions:** the logical focus for conservation of this species is Mt Kupe, where it is especially common and relatively well protected. Although reasonable data on population density is available, information is still lacking on regeneration levels and growth rates.

**SIMAROUBACEAE** (M. Cheek)

*Quassia sanguinea* Cheek & Jongkind
(syn.: *Hannoa ferruginea* Engl.)
VU A2c
**Range**: Nigeria (Obudu Plateau); Cameroon (SW, NW & W Regions).

Restricted to the submontane forests of the Cameroon Highlands, from Mt Cameroon in the S to the Bamenda Highlands in the N, with an extension into Nigeria at Obudu, this small tree, often only 3–4 m tall, is distinctive in its red leaf axes. Previously named as *Hannoa ferruginea*, justification for the name change and additional data on the species is given in: Cheek, M. & Jongkind, C. (2008). Two new names in West-Central African *Quassia* (Simaroubaceae). Kew Bull. 63: 247–250.

**Habitat**: submontane forest; 800–1750 m alt.

**Threats**: forest clearance for wood, followed by agriculture, particularly in the northern part of its range, the Bamboutos Mts and the Bamenda Highlands. In the latter, a remote sensing study over eight years (1987–1995) by Moat (in Cheek *et al.* 2000) of one area showed 25% loss of forest. By extrapolation, it is here estimated that over 30% of its overall population has been lost due to habitat destruction over the last three generations, or sixty years (estimating one generation at twenty years). **Management suggestions**: although forest loss in W and NW Regions has seriously reduced the population of *Q. sanguinea* in those areas, it seems relatively secure at Mt Cameroon and Mt Kupe-Bakossi Mts in SW Region. So long as these areas remain protected, no further action is needed to ensure the survival of the species. However, data on generation duration and other aspects of demography, together with data on densities, are desirable.

## STERCULIACEAE (M. Cheek)

*Leptonychia kamerunensis* Engl. & K. Krause
(syn.: *Leptonychia* **sp. 1 of Kupe-Bakossi checklist**; *Leptonychia* **sp. 1 of Bali Ngemba checklist)**.
EN B2a, b(iii)
**Range:** Cameroon endemic (SW: Mt Kupe, Lebialem Highlands; NW: Bali Ngemba FR).
This treatment is taken from Cheek in press (Three new or resurrected species of *Leptonychia* (Sterculiaceae-Byttneriaceae-Malvaceae) from West-Central Africa, Kew Bull.) in which the species is resurrected.

*Leptonychia kamerunensis* is the only member of the genus known to occur above 1000 m altitude W of the Albertine Rift that marks the eastern boundary of Congo (Kinshasa). It is further distinguished by having by far the largest fruits of any member of the genus. Most species have fruits only 1–1.5 cm diam. when mature. The local name at Fosimondi is Keliteh (Banwa language). The fruits can be stewed like those of okra (*Abelmoschus esculentus*) according to Nwopiemb and Tanjong of Fosimondi (*Tchiengué* 2208).

*Leptonychia kamerunensis* probably once occurred more or less continuously in the forests of the Cameroon Highlands from Mt Kupe in the S to the Bamenda Highlands in the N. Today it still has this range, but in the now largely deforested Bamenda Highlands it is restricted to the forest relic which is the Bali Ngemba FR. *L. kamerunensis* appears to be largely absent from the area in between. However it does survive just to the west of the Bamboutos Mts at Fosimondi. It is now only known at four sites: at Mt Kupe (above Kupe village and above Nyasoso); at Fosimondi; and at the Bali Ngemba FR. At the latter two it is under immediate and ongoing threat from clearance of the understorey for the cultivation of crops such as *Colocasia*. At Fosimondi, in addition, the forest canopy itself is also being progressively cleared downslope from the village of Fosimondi (c. 1900 m alt.), fide Tchiengué (pers. comm. 2006). The area of occupancy is estimated as being 50 km² (2007). At the site where it was first discovered, Mt Ekomane, SW of the the Bamboutos Mts, not far distant from Fosimondi between Mt Kupe and Bali Ngemba, forest clearance has been so extensive that the species is believed no longer to exist there. *L. kamerunensis* is here assessed as EN B2a,b (iii) by the criteria of IUCN (2001), that is endangered. *L. kamerunensis*, while not common, is not infrequent, at either Bali Ngemba (six specimens) or Mt Kupe (eight coll.). At Bali Ngemba it is listed as one of 17 taxa that characterise the submontane forest since collections occur there throughout the altitudinal range 1300 to 1900–2000 m alt. (Cheek in Harvey *et al.* 2004: 16). **Habitat:** submontane forest with *Santiria trimera* (Burseraceae), *Cola verticillata* (Sterculiaceae), *Macaranga occidentalis* (Euphorbiaceae); 1190– 2000 m alt.

**Threats:** clearance of surviving forest for timber, followed by agriculture especially at Fosimondi and Bali Ngemba (see above).

**Management suggestions:** at Fosimondi and Bali Ngemba the assistance of local communities should be sought and supported to protect this species and it should be incorporated into reforestation plantings. The logical focal point for conserving the species is at Mt Kupe where threats are lowest and the species has been most frequently recorded. Data on regeneration levels and growth rates are needed, as well as educational material to support its protection and management.

# MONOCOTYLEDONAE

## COMMELINACEAE (M. Cheek)

### *Aneilema silvaticum* Brenan
VU B2ab(iii)

**Range:** Nigeria; Cameroon; and Congo (Kinshasa).

Known at RBG Kew from three collections in Nigeria (*Meikle* 637, 1949; *J.D.Kennedy* 2674 & 1758, respectively, 1935 and 1931, all from Sapoba), three in Congo (Kinshasa) (*Lemaire* 379, Mobwassa; *Seret* 69, Uele; *Tilquin* 183, Kinshasa) and six in Cameroon (*Zenker* 1110, Bipinde, 1896 and *Mbatchou* 470, Mt Cameroon, 1992; also Lebialem-Bechati in SW; Chantier Sanaga in Littoral).

This is a tiny mat-forming herb in swamp forest, very likely to have been overlooked. It was first assessed in Cheek *et al.* 2004. That assessment is maintained here since although new sites have been added, the threshold of ten has not yet been exceeded. The ten sites equate to an AOO of 40 km$^2$ with 4 km$^2$ cells. Where the author has seen this species, it only occupied a site of about 4–5 m$^2$.

**Habitat:** lowland forest; 300–880 m alt.

**Threats:** clearance of lowland forest for timber and/or agriculture is known to be a cause for concern in Nigeria generally including Sapoba, but also at Mt Cameroon, Kinshasa and the Loum FR.

**Management suggestions:** better policing and higher protection of existing forest reserves would help secure the future of this species. Further intensive survey work is likely to yield additional sites leading to its transfer to NT status in future.

## ORCHIDACEAE (M. Cheek & B. J. Pollard)

### *Bulbophyllum nigericum* Summerh.
VU A2c, B2ab(iii)

**Range:** Nigeria (5 coll., at 5 sites); Cameroon (Mt Kupe (1 coll.); Bakossi Mts-Enyangdong, Lebialem, Bali Ngemba, Dom (1 coll. each), unlocated (2 coll.)).

This species was described from *King* 124 collected in October 1958 from Plateau Province, Nigeria, from where King made three additional collections. A further record, from the Mambilla Plateau in November 1993 (*Sporrier* 18) remains unconfirmed, the specimen being labelled "*Bulbophyllum nigericum*?". The specimen cited in FWTA from the Ivory Coast is now referred to *B. bidenticulatum* J.J.Vermeulen subsp. *bidenticulatum* (Vermeulen 1987: 167). It was first collected in Cameroon on the southern side of Mt Kupe in (?)1970, *Letouzey* 408, but has not since been recorded there. Two additional specimens are recorded from western Cameroon by Vermeulen (1987: 92), but no site locations are listed. Recent intensive surveys in Cameroon have revealed only two additional sites, at Enyandong in the Bakossi Mts (*Salazar* 6322, Oct. 2001) and at Bali Ngemba FR (*Plot voucher* BAL25, Apr. 2002). The foregoing assessment which appeared in Harvey *et al.* 2004, is maintained here since no new records have been added apart from that at Lebialem which does not take the species outside the VU threshold under criterion B. There are nine sites (AOO 36 km² with 4 km² cells).

**Habitat:** an epilith or epiphyte in submontane and montane forest; c. 800–2050 m alt.

**Threats:** the Nigerian sites are threatened by continued extensive clearance of forest to high elevations; one or more of these subpopulations are likely lost. The plant at Enyandong was found growing in largely cleared forest, on a tree in the village. It is therefore likely to occur in the surrounding forest, some of which is being cleared for small-holder farming, thus threatening this population. In all, a loss of over 30% of the population is estimated over the past three generations, which we here estimate to be 10 years, much of this loss being irreversible.

**Management suggestions:** as this species is found within the village of Enyandong, on a tree in front of the house of the Chief of the village, this is an ideal location for promoting community-based conservation. Local residents here could be encouraged to search for this species in the surrounding forest, perhaps using a species conservation poster as an aid to identification, and to promote protection of any locations where it is found.


### *Cyrtorchis letouzeyi* Szlach. & Olszewski

EN B2ab(iii)

**Range:** Cameroon endemic (SW: Lebialem; Centre: Ebolowa-Mbalomayo; E: S of Dja NP).

This epiphytic orchid has only tentatively been identified from Lebialem and its appearance is unexpected since it is distant from the other two locations and at much higher altitude. The species is assessed as EN here since only three locations are known (AOO 12 km² with 4 km² cells) and since threats are as below.

**Habitat:** evergreen forest; 545–1800 m alt.

**Threats:** at Lebialem, its habitat below Fosimondi village is highly threatened by clearance downslope for agriculture. Threats at the other sites are not known.

**Management suggestions:** this species should be established in cultivation and multiplied for reintroduction to the wild in protected areas. Better material of the taxon at Lebialem is needed to confirm its identity.


### *Diaphananthe bueae* (Schltr.) Schltr.

VU A2c + B2ab(iii)

Range: Cameroon endemic (Mt Cameroon (two pre-1988 collections), ?Banyang Mbo, Lebialem, ?Bali Ngemba and Bamenda Highlands (near Nkambe, Kumbo and at Dom)).

As the specific epithet suggests, this species was first collected at Buea on Mt Cameroon (*Deistel* s.n., type, collected 24 July 1905, "auf der altern Rinde hoher Baumen in Gesellschaft anderer Orchideen. Walden in d. Ungebung Buea"). It was rediscovered there over 40 years later (*Gregory* 153). More recently, in the 1970s, it has been found in the Bamenda Highlands (*Letouzey* 8889: Nkambe and Mbenkum; 354: Tadu-Kumbo). However, we were not able to discover this species at Mt Oku and the Ijim Ridge during 1996 or 1998. A possible record of this species from Ivory Coast (*Perez-Vera* 725, in fruit November 1974) needs support from flowering material before it is confirmed. The above assessment is taken from Cheek *et al.* 2000 where it was rated as EN A1c+2c. This assessment is downrated to VU under criterion B since potentially seven sites are now known (see list above, AOO 28 km$^2$ with 4 km$^2$ cells and threats as below) and under criterion A since we here estimate that more than 30% of its habitat has been lost in the last thirty years.

The identification of the species at Bali Ngemba is based on a field identification (Harvey *et al.* 2004). Despite relatively high sampling levels at Mt Kupe, the species was not picked up there (Cheek *et al.* 2004). The records from Banyang Mbo especially requires verification since the low altitude there is otherwise unknown in this taxon.

**Description:** epiphytic herb; leaves distichous, ligulate-oblong, 4–15 × 0.8–2 cm long, acute; inflorescence 6–15 cm long, 6–10-flowered. Flowers white and green. Sepals 6–8.5 × 2.7–2.8 mm. Petals oblong, 5–5.6 × 1.3–1.6 mm. Labellum narrowly or lanceolate ovate, 4–8.7 × 3.3–4.4 mm, apex acute, tooth longitudinally placed on the central vein; spur 12–15 mm long, apex swollen, emarginate. Stipites fused to apex, margins fimbriate; viscidium 0.5 mm long.

**Habitat:** submontane forest as an epiphyte; 1000–?1800 m alt.

**Threats:** all localities known are believed to be under pressure for forest clearance for agricultural, firewood collection and (e.g. Buea, Mt Cameroon where it may already be extinct) urban expansion.

**Management suggestions:** this species should be established in cultivation and multiplied for reintroduction to the wild in protected areas.


*Polystachya bicalcarata* Kraenzl.

VU A2c + 3c; B2ab(iii)

**Range:** Bioko (1 doubtful coll.); Cameroon (SW: Buea (3 coll.), Mt Cameroon unlocated (1 coll.), Mt Etinde (1 coll.), Banyang Mbo (1 coll.), Mt Kupe (2 coll. at 2 sites); W: Lebialem/Bamboutos Mts (2 coll.); NW: Mt Oku (1 coll.)).

Assessed in both Cable & Cheek (1998) and Cheek *et al.* (2000) as EN A1c + 2c, the assessment was altered in Cheek *et al.* 2004 in order to take into account the submontane sites on Mt Kupe, where this taxon's forest habitat remains largely intact, thus reducing the estimated percentage of habitat loss from >50% to >30%. This taxon is here also assessed under criterion B, being known from only 10 locations, one of which remains doubtful, with an area of occupancy of less than 500 km$^2$. The 2004 assessment is maintained here since the addition of the Banyang Mbo (Dr Zapfack) and Lebialem record does not take the taxon outside the existing rating under criterion B.

**Habitat:** an epiphyte of submontane and montane forest; 950–2000 m alt.

**Threats:** forest clearance for agriculture and firewood, particularly in the Bamboutos Mts and above Buea on Mt Cameroon.

**Management suggestions:** confirmation of this species' presence on Bioko is required. Surveys of this taxon at each of the collection localities should be made in order to assess its abundance. Continued protection of the submontane forest on Mt Kupe should ensure this species' future survival.

Fig. 7 *Nuxia congesta* (Buddlejaceae) by S. Ross-Craig.

W.E.T.

**FIG. 8** *Lobelia columnaris* (Campanulaceae) by W. E. Trevithick.

**FIG. 9** *Helichrysum cameroonense* (Compositae) by W. Fitch.

92

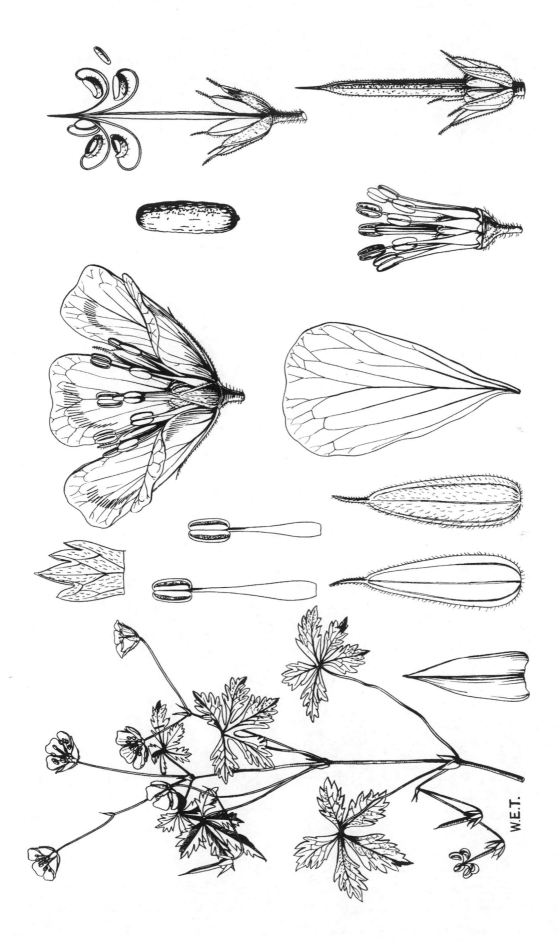

W.E.T.

FIG. 10 *Geranium arabicum* subsp. *arabicum* (Geraniaceae) by W.E. Trevithick.

93

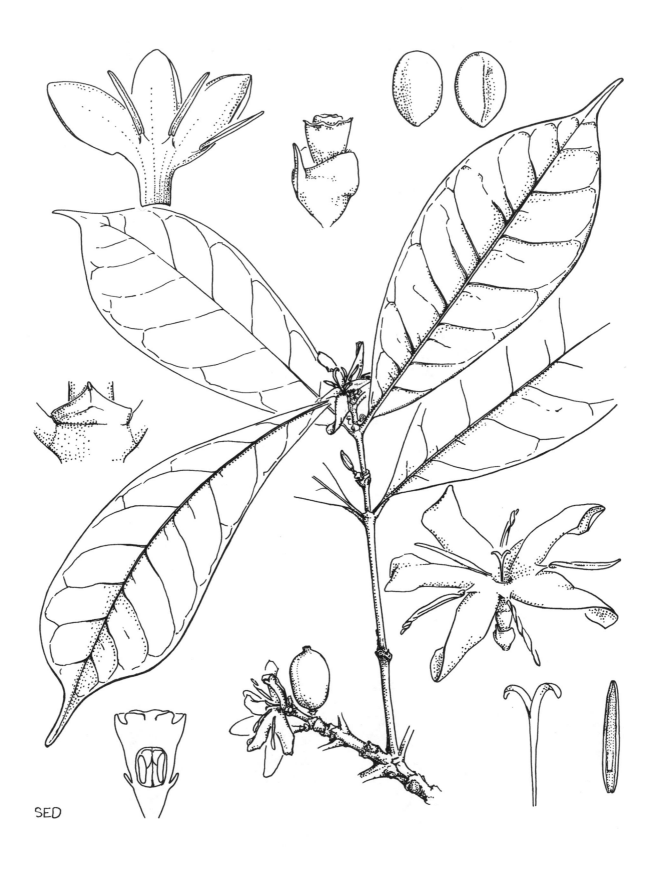

SED

**Fig. 11** *Coffea montekupensis* (Rubiaceae) by S. Dawson.

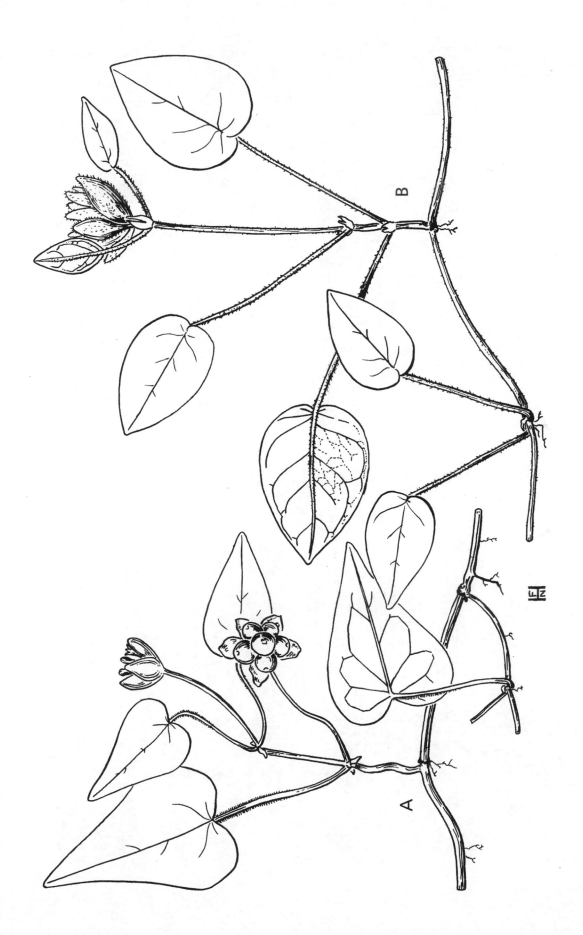

**FIG. 12** *Geophila afzelii* (B) & *G. obvallata* subsp. *obvallata* (A) (Rubiaceae) by F. N. Hepper.

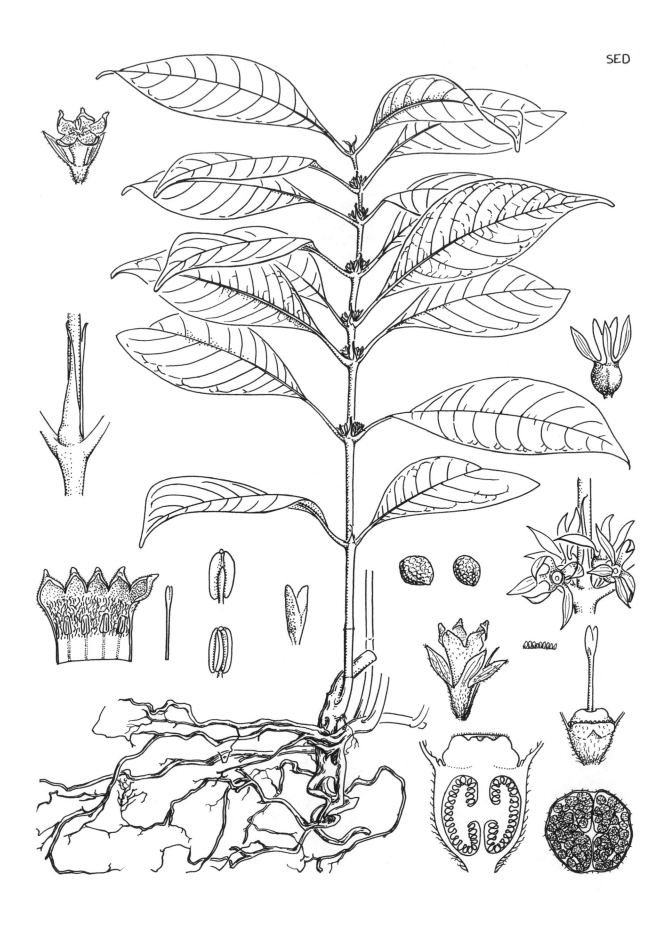

**Fig. 13** *Pentaloncha* sp. nov. of Kupe Bakossi checklist (Rubiaceae) by S. Dawson.

W.E.T.

**Fig. 14** *Clausena anisata* (Rutaceae) by W.E. Trevithick.

# BIBLIOGRAPHY

Berg, C.C., Hijman, M.E.E. & Weerdenburg, J.C.A.(1985). Moracées. *Flore Du Cameroun* 28. MESRES, Yaoundé.

Bergl, R.A., Oates, J.F. & Fotso, R. (2007). Distribution And Protected Area Coverage Of Endemic Taxa In West Africa's Biafran Forests And Highlands. *Biological Conservation* 134: 195–208.

Brummitt, R.K. (1992). *Vascular Plant Families and Genera.* Royal Botanic Gardens, Kew, U.K.

Brummitt, R.K. & Powell, C.E. (eds) (1992). *Authors of Plant Names.* Royal Botanic Gardens, Kew, U.K.

Cable, S. & Cheek, M. (eds) (1998). *The Plants of Mount Cameroon: A Conservation Checklist.* Royal Botanic Gardens, Kew, UK. lxxix + 198 pp.

Cheek. M., Harvey, Y. & Onana, J.-M. (eds) (2010). *The Plants of Dom, Bamenda Highlands, Cameroon: A Conservation Checklist.* Royal Botanic Gardens, Kew, UK. iv + 162 pp.

Cheek, M., Onana, J.-M. & Pollard, B.J. (eds) (2000). *The Plants of Mount Oku and the Ijim Ridge, Cameroon: A Conservation Checklist.* Royal Botanic Gardens, Kew, UK. iv + 211 pp.

Cheek, M., Pollard, B.J., Darbyshire, I., Onana, J.-M. & Wild, C. (eds) (2004). *The Plants of Kupe, Mwanenguba and the Bakossi Mountains, Cameroon: A Conservation Checklist.* Royal Botanic Gardens, Kew, UK. iv + 508 pp.

Convention on Biological Diversity (2002). *Convention on Biological Diversity: Text and Annexes.* United Nations Environment Programme, Montreal, Canada.

Cook, F.M. (1995). *Economic Botany Data Collection Standard.* RBG, Kew. 146 pp.

Courade, G. (1974). *Commentaire des cartes. Atlas Régional. Ouest I.* ORSTOM, Yaoundé.

Davies, A.P. & Sonké, B. (2009). A new *Argocoffeopsis* (Rubiaceae from Southern Cameroon: *Argocoffeopsis spathulata. Blumea* 53: 527–532.

de Ruiter, G. (1976). Revision for the genera *Myrianthus* and *Musanga* (Moraceae). *Bull. Jard. Bot. Nat. Belg.* 46: 471–510.

Harvey, Y., Pollard, B.J., Darbyshire, I., Onana, J.-M. & Cheek, M. (eds) (2004). *The Plants of Bali Ngemba Forest Reserve, Cameroon: A Conservation Checklist.* Royal Botanic Gardens, Kew, UK. iv + 154 pp.

Hawkins, P. & Brunt, M. (1965). *The Soils and Ecology of West Cameroon.* 2 Vols. FAO, Rome. 516 pp, numerous plates and maps.

Hepper, F.N. & Neate, F. (1971). *Plant collectors in West Africa : a biographical index of those who have collected herbarium material of flowering plants and ferns between Cape Verde and Lake Chad, and from the coast to the Sahara at 18° n.* International Bureau for Plant Taxonomy and Nomenclature. Utrecht: Oosthoek. xvi, 89, 8p. of plates.

IUCN (2001). *IUCN Red List Categories and Criteria. Version 3.1.* IUCN, Gland, Switzerland. 30 pp.

IUCN (2001). *IUCN Red List Categories and Criteria. Version 3.1.* IUCN, Gland, Switzerland. 30 pp.

IUCN (2003). *Guidelines for Application of IUCN Red List Criteria at Regional Levels: Version 3.0.* IUCN Species Survival Commission. IUCN, Gland, Switzerland and Cambridge, UK.

Keay, R.W.J. & Hepper, F.N. (eds) (1954–1972). *Flora of West Tropical Africa*, 2nd ed., 3 vols. Crown Agents, London.

Letouzey, R. (1968). Les Botanistes au Cameroun. *Flore du Cameroun 7.* Museum National d'Histoire Naturelle, Paris.

Letouzey, R. (1977). Presence de *Ternstroemia polypetala* Melchior (Theacees) dans les montagnes camerounaises. *Adansonia* 17 (1): 5–10.

Letouzey, R. (1985). *Notice de la carte phytogéographique du Cameroun au 1: 500,000.* IRA, Yaoundé, Cameroon.

Mabberley, D.J. (1997). *The Plant Book.* Second edition. Cambridge University Press, U.K.

Nkembi, L. & Atem, T. (2003) *A Report on the Biological and Socio-economic Activities conducted by the Lebialem Highlands Forest Project, South West Province, Cameroon.* Environment and Rural Development Foundation Report.

Sanderson, I.T. (1936). Amphibians of the Mamfe Division, Cameroon. *Proceed. Zool. Soc.* London. pp. 165–208.

van der Waarde, J. (2004). *Plant Collection Around Fossimondi, Bamboutos Mountains, South West Province, Cameroon, 27 and 28 October 2004.* Internal Project Report, project code 2004.022. 32 pp, November 2004.

Vermeulen, J.J. (1987). A taxonomic revision of the continental African *Bulbophyllinae. Orchid Monographs* 2. 300 pp, 101 figs, 11 plates.

A. Map signpost at Baranka/Mbelenka, April 2005, by Aline Horwath
B. Forest on scarp slope above Bechati, Sept. 2006, by Xander van der Burgt
C. Forest above Bechati en route for Fosimondi (with *Platycerium stemaria*), Sept. 2006, by Xander van der Burgt
D. Waterfall at Bechati, Sept. 2006, by Xander van der Burgt
E. Densely forested scarp slope between Bechati and Fosimondi, Sept. 2006, by Xander van der Burgt
F. View from Fosimondi looking West, agriculture in foreground, Feb. 2006, by Aline Horwath

**Plate 1**

A. Cloud forest, Fosimondi, Feb. 2006, by Barthélemy Tchiengué

B. April 2004 team near Fosimondi with Barthélemy Tchiengué, Nicole Guedje, Terence Atem, Terence Suwinyi, local guide and Kenneth Enow, by Nina Rønsted

C. Cloud forest with moss-festooned trees and the April 2005 team with Laura Pearce, Barthélemy Tchiengué & Dennis Ndeloh, by Aline Horwath

D. Agriculture with Eucalyptus trees above Fosimondi, Feb. 2006, by Barthélemy Tchiengué

E. Forest clearance in cloud forest, April 2005, by Aline Horwath

F. Forest being cleared for Banana farm, Feb. 2006, by Aline Horwath

G. Palm oil factory at Bechati, Sept. 2006, by Xander van der Burgt

**Plate 2**

A. *Justicia tenella* (Acanthaceae), Sept. 2006, Tchiengué 2755, by Xander van der Burgt

B. *Stenandrium guineense* (Acanthaceae), Sept. 2006, by Xander van der Burgt

C. *Thunbergia fasciculata* (Acanthaceae), Sept. 2006, van der Burgt 867, by Xander van der Burgt

D. *Monanthotaxis* sp. nov. of Bechati (Annonaceae), Sept. 2006, Tchiengué 2816, by Xander van der Burgt

E. *Schefflera hierniana* (Araliaceae), Feb. 2006, by Barthélemy Tchiengué

F. *Impatiens filicornu* (Balsaminaceae), April 2005, Tchiengué 2250, by Aline Horwath

G. *Impatiens kamerunensis* subsp. *kamerunensis* (Balsaminaceae), Sept. 2006, Tchiengué 2775, by Xander van der Burgt

**Plate 3**

A. *Impatiens letouzeyi* (Balsaminaceae), April 2005, Tchiengué 2236, by Aline Horwath
B. *Begonia pseudoviola* (Begoniaceae), Sept. 2006, van der Burgt 869, by Xander van der Burgt
C. *Lobelia columnaris* (Campanulaceae), Feb. 2006, by Barthélemy Tchiengué
D. *Salacia lebrunii* (Celastraceae) (flower), Sept. 2006, van der Burgt 870, by Xander van der Burgt
E. *Salacia lebrunii* (Celastraceae) (fruit), Sept. 2006, van der Burgt 870, by Xander van der Burgt
F. *Oncoba ovalis* (Flacourtiaceae), April 2005, by Aline Horwath
G. *Allanblackia gabonensis* (Guttiferae), Feb. 2006, by Barthélemy Tchiengué
H. *Psorospermum densipunctatum* (Guttiferae), Feb. 2006, Tchiengué 2561, by Aline Horwath

**Plate 4**

A. *Chlamydocarya thomsoniana* (Icacinaceae), Sept. 2006, van der Burgt 878, by Xander van der Burgt
B. *Pyrenacantha longirostrata* (Icacinaceae), Sept. 2006, van der Burgt 888, by Xander van der Burgt
C. *Dicot* indet. (field determination: Icacinaceae), Sept. 2006, by Xander van der Burgt
D. *Clerodendrum globuliflorum* (Labiatae), Feb. 2006, Tchiengué 2589, by Barthélemy Tchiengué
E. *Clerodendrum silvanum* var. *buchholzii* (Labiatae), Sept. 2006, van der Burgt 879, by Xander van der Burgt
F. *Plectranthus insignis* (Labiatae), April 2005, Tchiengué 2223, by Aline Horwath
G. *Napoleonaea egertonii* (Lecythidaceae), Sept. 2006, by Xander van der Burgt

**Plate 5**

A. *Duparquetia orchidacea* (Leguminosae-Caesalpinioideae) (flower), Sept. 2006, van der Burgt 862, by Xander van der Burgt
B. *Duparquetia orchidacea* (Leguminosae-Caesalpinioideae) (fruit), Sept. 2006, van der Burgt 862, by Xander van der Burgt
C. *Cylicodiscus gabunensis* (Leguminosae-Mimosoideae), Sept. 2006, photographic record by Xander van der Burgt
D. *Medinilla mirabilis* (Melastomataceae), Sept. 2006, van der Burgt 892, by Xander van der Burgt
E. *Heckeldora staudtii* (Meliaceae), Sept. 2006, van der Burgt 883, by Xander van der Burgt
F. *Bersama abyssinica* (Melianthaceae), Sept. 2006, van der Burgt 876, by Xander van der Burgt
G. *Penianthus longifolius* (Menispermaceae), April 2005, by Aline Horwath
H. *Perichasma laetificata* var. *laetificata* (Menispermaceae), Sept. 2006, van der Burgt 864, by Xander van der Burgt

**Plate 6**

A. *Argocoffeopsis fosimondi* (Rubicaceae), Feb. 2006, by Aline Horwath
B. *Chassalia laikomensis* (Rubiaceae), April 2005, Tchiengué 1914, by Aline Horwath
C. *Coffea montekupensis* (Rubiaceae), Feb. 2006, Tchiengué 2617 by Barthélemy Tchiengué
D. *Geophila obvallata* subsp. *obvallata* (Rubiaceae), Sept. 2006, van der Burgt 851, by Xander van der Burgt
E. *Heinsia crinita* (Rubiaceae), Sept. 2006, van der Burgt 882, by Xander van der Burgt
F. *Hymenocoleus subipecacuanha* (Rubiaceae), Sept. 2006, Tchiengué 2837, by Xander van der Burgt
G. *Mussaenda* cf. *tenuiflora* (Rubiaceae), Sept. 2006, Tchiengué 2798, by Xander van der Burgt

**Plate 7**

A. *Zanthoxylum rubescens* (Rutaceae), April 2005, by Aline Horwath
B. *Rhaptopetalum geophylax* (Scytopetalaceae), Tchiengué 2582, Feb. 2006, by Aline Horwath
C. *Microcos barombiensis* (Tiliaceae), Sept. 2006, van der Burgt 895, by Xander van der Burgt
D. *Coleotrype laurentii* (Commelinaceae), Sept. 2006, van der Burgt 891, by Xander van der Burgt
E. *Costus letestui* (Costaceae), Feb 2006, Tchiengué 2592, by Aline Horwath
F. *Eulophia horsfallii* (Orchidaceae), Feb. 2006, by Aline Horwath
G. *Habenaria weileriana* (Orchidaceae), Sept. 2006, van der Burgt 852, by Xander van der Burgt
H. *Selaginella vogelii* (Selaginellaceae), April 2005, by Aline Horwath
I. *Cyathea camerooniana* var. *camerooniana* (Cyatheaceae), Feb. 2006, by Aline Horwath

**Plate 8**

# READ THIS FIRST:
# EXPLANATORY NOTES TO THE CHECKLIST

Yvette Harvey

Herbarium, Royal Botanic Gardens, Kew, Richmond, Surrey, TW9 3AE, UK

Before using this checklist, the following explanatory notes to the conventions and format used should be read.

The checklist is compiled in an alphabetical arrangement: species within genera, genera within families and families within the groups Dicotyledonae, Monocotyledonae, Pinopsida, Lycopsida (fern allies) and Filicopsida (true ferns), following Kubitzki, in Mabberley (1997: 771–781).

Identifications and descriptions of the species were carried out on a family-by-family basis by both family specialists and general taxonomists; these authors are credited at the head of each account. As a general rule, if two authors are listed, the primary author is responsible for the determinations and the second author for the compiling of the account, including writing of descriptions, distributional data and conservation assessments.

As the incomplete Flore du Cameroun (1963–) is a particularly relevant source of information on the plants of the checklist area, a reference to the volume and year of publication is listed at the head of each family account where available.

The families and genera accepted here follow Brummitt (1992), with recent updates on the Kew Vascular Plant Families and Genera database. Within the checklist each taxon in the family account is treated in the following manner:

## TAXON NAME

The species name adopted follows the most recent taxonomic work available. Author abbreviations follow the standards of Brummitt & Powell (1992) (www.ipni.org).

Species names not validly published at the time of publication of the checklist are noted as "in press" or "ined." depending upon the extent to which the publication process has advanced, the former indicating that the protologue has been accepted for, but is awaiting, publication and the latter that the species concept is firmly established but that the protologue is not yet submitted for publication.

Not all names listed are straightforward binomials with authorities. A generic name followed by "sp." generally indicates that the material was inadequate to name to species, for example *Crotalaria* sp. (Leguminosae: Papilionoideae). Use of "sp. 1", "sp. 2" and "sp. A", "sp. B" etc. generally indicate unmatched specimens which may be new to science or may prove to be variants of a currently accepted species; these taxa usually require additional material in order to confirm identity, for example *Psychotria* sp. B of Bali Ngemba Checklist (Rubiaceae), or new taxa for which sufficient material is available, but are awaiting formal description, for example *Voacanga* sp. 1 of Bali

Ngemba & Kupe (Apocynaceae). Unless otherwise stated, or inferred from the distribution, these provisional names are applicable only to the current checklist, thus "sp. 1 of Fosimondi" or "sp. 1 of Bechati" indicates "sp. 1 of the Lebialem Highlands, Cameroon checklist". The use of "sp. nov." is a firm statement that the taxon is new to science but awaiting formal description; sufficient material may or may not be available for this process. A generic name followed by "cf.": indicates that the specimens cited should be compared with the associated specific epithet, for example *Tchiengué* 2762 (*Microdesmis* cf. *puberula* Hook.f.) should be compared with *Microdesmis puberula* Hook.f. This is an indication of doubt (sometimes due to poor material), suggesting that the taxon is close to (but may differ from) the described taxon. The terms "aff.", indicating that the taxon has affinity to the subsequent specific epithet, and "vel. sp. aff.", indicating that the specimen refers to the taxon listed or a closely allied entity, are applied in a similar fashion. These uncertainties are generally explained in the taxon's "Notes" section (see below).

## TAXON REFERENCE

The majority of species referred to within the checklist are found in the 2nd edition of Flora of West Tropical Africa (FWTA: Keay 1954–58; Hepper 1963–72), the standard regional flora. Only species names which do not occur in FWTA are given a reference here; if no reference is cited the taxon name is currently accepted and occurs in FWTA. The references listed are not necessarily the place of first publication of the name; rather, we have tried where possible to use widely accessible publications which provide useful information on that taxon, such as a description and/or distribution and habitat data. The reference is recorded immediately below the taxon name. In the case of scientific journals, we list the journal name, volume and page numbers and date of publication, with recording of volume part number where it aids in access to the publication. With the exception of our series of Conservation Checklists, in the case of books, we list the surname of the author(s), the book title, the page number for the taxon in question and the year of publication. Journal and book titles are often abbreviated in the interest of economy of space. Several notable publications are:

**Fl. Cameroun**
Flore du Cameroun (1963–) Muséum National d'Histoire Naturelle, Paris, France & MINREST, Yaoundé, Cameroon

**Fl. Gabon**
Flore du Gabon (1961–) Muséum National d'Histoire Naturelle, Paris, France.

**F.T.E.A.**
The Flora of Tropical East Africa (1952–) Crown Agents, London & A.A.Balkema, Lisse, Netherlands.

## SYNONYMS

In some instances, names used in FWTA have been superseded and are thus reduced to synonymy; these are listed below the accepted name, with the prefix "Syn.". Names listed in synonymy in FWTA are not recorded here. Other important synonyms are, however, recorded.

## TAXON DESCRIPTION

The short descriptions provided for each taxon are based primarily upon the material cited in order to provide the most accurate representation for field botanists working within the checklist area. However, where necessary, they are supplemented by extracted details from the descriptions in FWTA, Fl. Cameroun and the cited taxonomic works. The descriptions are not exhaustive or necessarily diagnostic; rather, they aim to list the key characters to enable field identification of live or dried material, thus microscopic or complex characters are referred to only when they are essential for identification. Where two or more taxa closely resemble one another, a comparative description may be used, by for example stating "…herb… resembling *Impatiens filicornu*, but …"; such comparisons are only made to other taxa occurring within the checklist area.

Several abbreviations are used in the descriptions, most notably d.b.h. (referring to "diameter at breast height", being a standard measure of the diameter of a tree trunk), the use of "c." as an abbreviation for "approximately", "±" meaning "more or less".

## HABITAT

The habitat, recorded at the end of the description, is derived mainly from the field notes of the cited specimens and therefore does not necessarily reflect the entire range of habitats for that taxon; rather, those in which it has been recorded within the checklist area. Habitat information is taken from published sources only where field data is not available, for example, where the only specimens recorded were not available to us, but are cited in FWTA. Altitudinal measurements are derived from barometric altimeters carried by the collectors. Altitudinal ranges, listed together with habitat, are generated directly from the database of specimens from the checklist area, and thus do not necessarily reflect the entire altitudinal range known for the taxon. Where no altitudinal data was recorded with the specimens, it is omitted.

## DISTRIBUTION

For the sake of brevity, country ranges are generally recorded for each taxon rather than listing each separate country, for example "Sierra Leone to Uganda" is taken to include all or most of the intervening countries. Only where taxa are recorded from only two or three, rarely more, countries within a wide area of occurrence are the individual countries listed, for example "Sierra Leone, Cameroon & Uganda". For more widespread taxa, a more general distribution such as "tropical Africa" or "pantropical" is recorded. Where a species is alien to the checklist area, its place of origin is noted, together with its current distribution. Several country abbreviations are used:

Guinea (Bissau):        former Portuguese Guinea

Guinea (Conakry):    the Republic of Guinea, or former French Guinea

CAR:                          Central African Republic

Congo (Brazzaville):  the Republic of Congo or former French Congo

Congo (Kinshasa):     the Democratic Republic of Congo, or former Zaïre

Abbreviations for parts of the country are also used (N: north, S: south (or southern), E: east, W: west, C: central), with the exception of South Africa. Where appropriate, Equatorial Guinea is divided into Bioko, Annobón (both islands) and Rio Muni (mainland), and the Angolan enclave of Cabinda North of the Congo River is recorded separately from Angola itself (south of the river).

In addition to country range, a chorology, largely based upon the phytochoria of White (1983) but with modifications to reflect localised centres of endemism in W Africa, is recorded in square brackets. The main phytochoria used are:

**upper Guinea**
Broadly the humid zone following the Guinean coast from Senegal to Ghana.

**lower Guinea**
Separated from upper Guinea by the "Dahomey Gap", an area of drier savanna-type vegetation, that reaches the atlantic coast. Lower Guinea represents the humid zone from Nigeria to Gabon, including Rio Muni, Cabinda, and the wetter parts of western Congo (Brazzaville).

**Congolian**
The basin of the River Congo and its tributaries, from eastern Congo (Brazzaville) and southern CAR, through Congo (Kinshasa) and to Uganda, Zambia and Angola.

**afromontane**
A series of vegetation types restricted to montane regions, principally over 2000 m alt.

**W(estern) Cameroon Uplands**
A subdivision of the Afromontane phytochorion, used for taxa restricted to the mountain chain running from the Gulf of Guinea islands (Annobón, São Tomé, Príncipe and Bioko) to western Cameroon and south-east Nigeria.

**Cameroon endemic**
For those taxa restricted to Cameroon, a subdivision of lower Guinea. Taxa endemic to montane western Cameroon are however recorded under W Cameroon Uplands unless they are endemic to the checklist area, when they are listed as a narrow endemic (see below).

**narrow endemic**
For those taxa restricted to the checklist area, a subdivision of W Cameroon Uplands.

These phytochoria are variously combined where appropriate, for example "Guineo-Congolian (montane)" refers to an afromontane taxa restricted to the mountains of the upper and lower Guinea and Congolian phytochoria. Taxa with ranges largely confined to the Guineo-Congolian phytochorion, but with small outlying populations in wet forest in, for example, west Tanzania or northern Zimbabwe, are here recorded as Guineo-Congolian rather than tropical African, as the latter would provide a more misleading representation of the taxon's true phytogeography. A range of other chorologies are used for more widespread species, including "tropical Africa", "tropical & S Africa", "tropical Africa & Madagascar", "palaeotropics" (taxa from tropical Africa and Asia or some other Old World region), "amphi-Atlantic" (taxa from tropical Africa and S America), "pantropical" and "cosmopolitan". If these terms are used in the distribution, no separate phytochorion is listed.

For some taxa, such as those native to one area of the tropics but widely cultivated elsewhere, the chorology is difficult to define and is thus omitted. Both distribution and chorology are omitted for taxa where there is uncertainty over its identification.

## CONSERVATION ASSESSMENT

The level of threat of future extinction on a global basis is assessed for each taxon that has been fully identified and has a published name, or for which publication is imminent, under the guidelines of the IUCN (2001). Under the heading "IUCN:", each taxon is accredited one of the following Red List categories:

LC:    Least Concern

NT:    Near-threatened

VU:    Vulnerable

EN:    Endangered

CR:    Critically Endangered

Those taxa listed as VU, EN or CR are treated in full within the chapter on Red Data species, where the criteria for assessment are recorded. Those listed as LC or NT are not treated further in this publication, but it is recommended that further investigation of the threats to those taxa recorded as NT are made. Undescribed taxa, or those with an uncertain determination, are not assessed. In addition, we do not consider it appropriate to assess taxa from the poorly-known genus *Anchomanes* (Araceae) for which species delimitation is currently poorly understood and thus for which conservation assessments would be somewhat meaningless; this genus should be revisited once a full taxonomic revision has been completed.

## SPECIMEN CITATIONS

Specimens from the checklist area are recorded below the distribution and conservation assessment, with the following information:

*Location*: Fosimondi and Bechati are a rough indicators as to the actual location, which can be provided by reference to the cited specimen.

*Collector and number*: within each location, collectors are ordered alphabetically and, together with the unique collection number, are underlined. Only the principal collector is listed here, thus for example, some of the collections listed under "Tchiengue" may originally have been recorded as "Tchiengue, Horwath, Pearce, Ndeloh, Nuopiemba".

*Phenology*: this information is derived directly from the Cameroon specimens database at Kew and is thus dependent upon recording of such information at the time of collection or at the point of data entry onto the database; if this was not done no phenological information is listed. Collections are recorded as flowering (fl.), fruiting (fr.) or sterile (st.) where applicable.

*Date*: the month and year of collection is recorded for each specimen; within each collector from each location, collections are ordered chronologically.

The specimens cited are of herbarium material derived from a variety of sources; the chapter on collectors in the checklist area provides further information. In addition, a very few confident sight records of uncollected taxa are also included, also photographic records where no specimen was obtained.

## NOTES

Items recorded in the notes field at the end of the taxon account include:

Taxonomic notes, for example in the case of new or uncertain taxa, how they differ from closely related species.

Notes on the source of specimen data, for example for those specimens recorded in Fl. Cameroun, or for specimens not seen by the author(s) of the family account.

## ETHNOBOTANICAL INFORMATION

Local names and uses are listed for each taxon where appropriate; these are derived largely from local residents and field assistants and are reproduced here with their consent. Local names are in the Banwa language unless otherwise stated. For each local name or use listed, a source is attributed, usually by reference to the collector and number of the specimen from which the information was derived. The sources of the name or use, where recorded, were the local guides who accompanied the teams. In circumstances where the ethnobotanical information was provided verbally with no attached specimen, or where it is known by the author(s) of the family account, the terms *"fide"*, *"pers. comm."* (personal communication) or *"pers. obs."* (personal observation) are used to attribute the information to a source.

The layout of the information on local uses follows the convention of Cook (1995), listing level 1 state categories of use in capital letters, followed by the level 2 state categories, then the specific use. In order to comply with guidelines set by the Convention on Biological Diversity (2002), the detailed uses of medicinal plants are not listed here; only the level 1 state "MEDICINES" is recorded.

## REFERENCES

Brummitt, R.K. (1992). *Vascular Plant Families and Genera.* Royal Botanic Gardens, Kew, U.K.

Brummitt, R.K. & Powell, C.E. (eds) (1992). *Authors of Plant Names.* Royal Botanic Gardens, Kew, U.K.

Convention on Biological Diversity (2002). *Convention on Biological Diversity: Text and Annexes.* United Nations Environment Programme, Montreal, Canada.

Cook, F.M. (1995). *Economic Botany Data Collection Standard.* RBG, Kew. 146 pp.

IUCN (2001). *IUCN Red List Categories and Criteria. Version 3.1.* IUCN, Gland, Switzerland. 30 pp.

Keay, R.W.J. & Hepper, F.N. (eds) (1954–1972). *Flora of West Tropical Africa*, 2nd ed., 3 vols. Crown Agents, London.

Mabberley, D.J. (1997). *The Plant Book.* Second edition. Cambridge University Press, U.K.

# ANGIOSPERMAE

## DICOTYLEDONAE

### ACANTHACEAE

I. Darbyshire (K)

**Brachystephanus giganteus** Champl.
Syst. Geogr. Pl. 79: 139 (2009).
**Syn.** *Oreacanthus mannii* Benth.
Mass-flowering herb or subshrub, 1–4m; stems with two opposite rows of hairs when young, glabrescent; leaves ovate to elliptic, 10–34 × 3–13cm, base cuneate-attenuate, apex acuminate, margin subcrenulate, surfaces sparsely pubescent, principally along veins, cystoliths present; inflorescence a many-branched panicle 13–28 × 5–12cm, usually densely glandular hairy; bracts and bracteoles linear to 5–15mm; calyx lobes subulate, 1.5–3.5mm; corolla 2-lipped, glabrous, tube short, 1–2mm, white, lips blue to purple, upper lip lanceolate, 5–8.5mm long, lower lip oblong, 5–8.5mm long, 3-lobed; stamens 2, exserted, anthers monothecous; capsule stipitate, 13–18mm long, glabrous, 4-seeded. Forest and forest margins; 1350–1540m.
**Distr.:** Cameroon & Bioko [W Cameroon Uplands].
**IUCN:** VU
**Fosimondi:** Tchiengue, 2564 2/2006; 2596 2/2006.

**Brachystephanus jaundensis** Lindau subsp.
**jaundensis**
Kew Bull. 51: 760 (1996); Syst. Geogr. Pl. 79: 167 (2009).
Perennial herb or subshrub, to 80cm; stems pubescent or subglabrous; leaves ovate-elliptic, 8.5–18 × 3.7–7cm, base attenuate, apex acuminate, margin subentire, largely glabrous or nerves beneath puberulous, cystoliths present; petiole to 5cm; inflorescence terminal, spiciform, to 15cm long, many-flowered, usually glandular hairy; bracts linear-lanceolate, to 9mm long; calyx lobes linear 6–12mm long; corolla 2-lipped, puberulous, tube narrowly cylindrical, 30–45mm, limb pink to purple, lips 7.5–10.5mm long, lower lip minutely 3-lobed; stamens 2, exserted, anthers monothecous; capsule stipitate, 10–14.5mm long, glabrous, 4-seeded. Forest; 1355m.
**Distr.:** Bioko, Cameroon, Gabon & Congo (Brazzaville) [lower Guinea].
**IUCN:** NT
**Fosimondi:** Tchiengue, 2610 2/2006.

**Brillantaisia vogeliana** (Nees) Benth.
Annual or perennial herb, to 1.5m; stems largely glabrous; leaves broadly ovate, 7–20 × 5 × 12cm, base truncate to subcordate, margin dentate, apex (sub)acuminate, hairy mainly along principal veins, cystoliths present; petiole to 10cm, upper portion often with irregularly toothed wing; panicle lax, 5–20cm long, many-flowered, glandular hairy; main axis bracts ovate, leafy, reducing upwards, sometimes caduceus; calyx lobes linear-lanceolate ± 1cm long; corolla 2-lipped, blue or purple, tube white 5–7mm long, lips 10–18mm long, upper lip hooded, lower lip shortly 3-lobed; stamens 2, held in upper lip, bithecous; capsule linear, 17–23mm long, over 10-seeded. Forest & forest margins; 1400m.
**Distr.:** Ghana to Congo (Kinshasa), Sudan to Kenya [Guineo-Congolian].
**IUCN:** LC
**Fosimondi:** Tchiengue, 2566 2/2006.
Note: Tchiengue 2566 is tentatively placed in this taxon.

**Dischistocalyx thunbergiiflorus** (T.Anderson) Benth. ex C.B.Clarke
Epiphytic or terrestrial herb or subshrub, to 2m; stems glabrous; leaves oblong-elliptic or somewhat obovate, 12–27 × 6.5–8cm, pairs unequal, base acute, apex acuminate, margin crenulate, lower surface often reddish, glabrous; cystoliths present; petiole to 2.5cm; inflorescence terminal, spiciform or few-flowered, to 10cm; bracts early-caducous; calyx lobes lanceolate, 2–3cm long, upper 3 lobes partially fused; corolla funnel-shaped with spreading subregular limb, 5–7.5cm long, lobes blue or purple; stamens 4, bithecous; capsule c. 2–2.8cm, over 10-seeded. Forest including along streams; 1300–1340m.
**Distr.:** Cameroon & Bioko [W Cameroon Uplands].
**IUCN:** LC
**Fosimondi:** Tah 322 4/2004; Tchiengue, 2587 2/2006.
Note: *Dischistocalyx* is in much need of revision but it appears that specimens determined as *D. grandifolius* C.B.Clarke in the checklist of the Plants of Kupe, Mwanenguba & the Bakossi Mountains should have been named *D. thunbergiiflorus* (Darbyshire, viii.2009).

**Graptophyllum glandulosum** Turrill
Herb or shrub, to 2.5m; stems robust, woody at base, finely pubescent or glabrescent; leaves elliptic, 14–19 × 6.5–9cm, base acute, apex acuminate, margin subcrenulate, glabrous, cystoliths present; petiole c. 2cm; inflorescence terminal, racemoid, to 5.5cm, glandular-pubescent; bracts lanceolate to 4mm long; pedicels c. 0.7cm; calyx lobes subulate c. 4mm; corolla 3 × 0.7cm, 2-lipped, orange-red; stamens 2, orange, exserted, bithecous. Forest; 1320m.
**Distr.:** SE Nigeria & Cameroon [lower Guinea].
**IUCN:** NT
**Fosimondi:** Tchiengue, 2620 2/2006.

**Justicia tenella** (Nees) T.Anderson
Small creeping or decumbent herb, to 20cm; stems white-pubescent; leaves ovate, c. 2 × 1.3cm, base obtuse, apex acute or obtuse, margin subentire, largely glabrous, cystoliths present; petiole c. 1.3cm; inflorescence a dense spike 1–2cm long, lateral in clusters of 2–3, peduncle to 1.7cm; bracts obovate-orbicular, 2.5mm diam., pale green with paler margin,

imbricate; corolla minute, 2-lipped, lower lip purple with white margin; stamens 2, thecae offset, the lower tailed; capsule shortly stipitate c. 3mm long, glabrous, 4-seeded. Farmbush, grassland and forest margins; 250m.
**Distr.:** tropical Africa & Madagascar [tropical Africa & Madagascar].
**IUCN:** LC
**Bechati:** Tchiengue, 2755 9/2006.

### *Justicia* sp. of Fosimondi
Herb, to 40cm; stem with 2 lines of hairs; bracts green, whitish on sides. Forest understorey; 1340m.
**Fosimondi:** Tchiengue, 2604 2/2006.
Note: Tchiengue 2604 not seen by ID; a unicate at YA, identified as a *Rungia* (= *Justicia*) by Tchiengue, x.2008.

### *Mimulopsis solmsii* Schweinf.
Mass-flowering herb, to 1(–4)m; stems 4-angled, glabrescent; leaves ovate, 8–14 × 5–9cm, base shallowly cordate, apex acuminate, margin coarsely dentate, largely glabrous, cystoliths present; petiole to 5.5cm, puberulous; inflorescence a large terminal panicle, glandular pubescent; bracts elliptic, 1.3 × 0.3cm; bracteoles linear; calyx lobes linear to 1.3cm, glandular pubescent; corolla tube funnel-shaped, to 2.5cm long, pale pink or mauve, limb subregular; stamens 4, bithecous; capsule 15–25mm long, hairy, 6–8-seeded. Forest & forest edge; 1370–1460m.
**Distr.:** Bioko & Cameroon to E Africa [afromontane].
**IUCN:** LC
**Fosimondi:** Tchiengue, 2279 4/2005; 2602 2/2006.

### *Pseuderanthemum ludovicianum* (Büttner) Lindau
Erect herb or shrub, to 2m; stems robust, glabrous; leaves elliptic, 13.5–28 × 4.2–15cm, base acute, apex shortly acuminate, margin subcrenulate, cystoliths present; petiole to 2.5cm; inflorescence a terminal thyrse to 14cm with peduncle to 9cm; cymules closely spaced and many-flowered; bracts inconspicuous; calyx lobes lanceolate c. 3mm long; corolla 2-lipped, tube narrowly cylindrical, c. 35 × 1.5mm, upper lip 2-lobed, reflexed, lower lip 3-lobed, white speckled purple, puberulent; stamens 2, barely exserted, bithecous; capsule stipitate, to 3cm long, 4-seeded. Forest; 1370m.
**Distr.:** Liberia to Cameroon & S Sudan to NW Tanzania [Guineo-Congolian].
**IUCN:** LC
**Fosimondi:** Tchiengue, 2603 2/2006.

### *Stenandrium guineense* (Nees) Vollesen
Kew Bull. 47: 182 (1992).
**Syn.** *Stenandriopsis guineensis* (Nees) Benoist
Fl. Gabon 13: 102 (1966).
**Syn.** *Crossandra guineensis* Nees
Stoloniferous herb, to 50cm; stems tomentose; leaves in rosettes, elliptic, 9–14.5 × 4.5–5.7cm, base abruptly rounded or subordate, apex obtuse, margin subentire, blade pale green or purplish below, dark green above, variegated when young, veins pubescent beneath,

cystoliths absent; petiole to 2.5cm, tomentose; inflorescence a narrow terminal spike to 13cm long; bracts oblanceolate-rhombic, 1.7 × 0.6cm, apex mucronate, uppermost third with a regularly serrulate margin, striate, pink-tinged; calyx lobes subulate to 7mm long; corolla with a linear tube to 2cm, lobes spreading, pale purple with darker markings in throat; stamens 4, included, monothecous; capsule to 11mm long, glabrous, 4-seeded. Forest; 320m.
**Distr.:** Guinea (Conakry) to Angola (Cabinda) & Uganda to Sudan [Guineo-Congolian].
**IUCN:** LC
**Bechati:** Tchiengue, 2778 9/2006; 2779 9/2006; van der Waarde, 1 10/2004; 11 10/2004.

### *Thunbergia fasciculata* Lindau
Herbaceous twiner, to 8m; stems puberulous at nodes; leaves ovate, 7–9.5 × 5–7cm, base cordate, apex acuminate, margin irregularly undulate with apiculate vein apices, surfaces puberulous, cystoliths absent; petiole to 9cm; axillary inflorescence 2–3 flowered, peduncle to 12.5cm, puberulous; bracts paired, leafy, cordiform, 5.5 × 4cm, margin dentate, bracteoles elliptic, 2.5cm long, mucronate; corolla trumpet-shaped, to 5cm long, purple with yellow throat; stamens 4, included, bithecous; capsule woody, with prominent apical beak, pilose. Forest; 400m.
**Distr.:** Togo to Congo (Kinshasa), Sudan & SW Ethiopia, Uganda & W Kenya [Guineo-Congolian].
**IUCN:** LC
**Bechati:** van der Burgt, 867 9/2006; van der Waarde, 7 10/2004.

# ALANGIACEAE

## B. Tchiengué (YA)

Fl. Cameroun 10 (1970).

### *Alangium chinense* (Lour.) Harms
Tree, 5–18m, glabrescent; leaves alternate, ovate, c. 10 × 6cm, acuminate, obliquely rounded, entire, digitately 5-nerved, scalariform; inflorescences cymose, axillary, c. 20-flowered, 4cm; peduncle 1.5cm; flowers orange, 1.1cm; fruit ellipsoid, 1cm, fleshy. Forest margins and farmbush; 1500m.
**Distr.:** palaeotropical, excluding upper Guinea.
**IUCN:** LC
**Fosimondi:** Tchiengue, 2273 4/2005.

# AMARANTHACEAE

## C.C. Townsend (K), B. Tchiengué (YA) & Y.B. Harvey (K)

Fl. Cameroun 17 (1974).

### *Celosia isertii* C.C.Towns.
F.T.E.A. Amaranthaceae: 15 (1985); Fl. Zamb. 9(1): 34 (1988).

Climbing or scrambling herb, to 2m; stems glabrous; leaves drying green-black, alternate, ovate, 6.8–9.3 × 3–5cm, apex acuminate-apiculate, base attenuate then abruptly truncate, margin irregularly undulate, glabrous or scabridulous, veins occasionally purplish below; petiole c. 2.5cm; inflorescence spikes axillary or terminal, occasionally branched, 2.5–10cm; peduncle to 3.5cm, glabrescent; bracteoles and tepals ovate-elliptic, white, the latter 3–4mm long; style bifid. Forest & forest edge; 590m.
**Distr.:** tropical African highlands [afromontane].
**IUCN:** LC
**Bechati:** Tchiengue, 2825 9/2006.

### Celosia leptostachya Benth.
Suberect branching herb, to 40cm; stems furrowed, glabrous; leaves alternate, ovate-deltate, 3.4–5.6 × 2.3–3.8cm, apex acute, base attenuate then abruptly truncate, margin irregularly undulate, glabrous; petiole c. 2cm; spikes terminal, 8–23cm; peduncle fine, glabrous; flowers in laxly arranged glomerules, c. 6-flowered, green-brown when dry; bracteoles and tepals ovate, c. 2.5mm long; style bifid. Forest; 500m.
**Distr.:** Nigeria to Congo (Kinshasa) [lower Guinea & Congolian].
**IUCN:** LC
**Bechati:** van der Burgt, 889 9/2006.

### Celosia pseudovirgata Schinz
Fl. Cameroun 17: 13 (1974).
**Syn. Celosia bonnivairii** *sensu* Keay
Suberect herb, to 0.4–1m; stems glabrescent or finely pubescent; leaves alternate, ovate, 4.5–10(–14.5) × 2.5–7cm, acuminate, base shortly attenuate, margin subentire to obscurely undulate; petiole 1–3.5cm; inflorescence spikes terminal or occasionally axillary, sometimes branched, 2.5–5cm; peduncle to 4cm, glabrescent or finely pubescent; bracteoles and tepals brown; tepals ovate, c. 2mm. Forest & forest edges; 1350–1800m.
**Distr.:** Nigeria to Congo (Kinshasa) [lower Guinea & Congolian].
**IUCN:** LC
**Fosimondi:** Tchiengue, 1967 4/2004; 2599 2/2006.

## ANACARDIACEAE

### Y.B. Harvey (K)

### Antrocaryon klaineanum Pierre
Tree, 15m, no exudate; leaves pinnate, 30–40cm, 8-jugate; leaflets elliptic c. 6 × 2.5cm, subacuminate, obtuse, nerves 5 pairs; petiolule 0.5cm; petiole 13cm; inflorescence spike-like, 18cm, supra-axillary; flowers numerous, 1–2mm; fruit top-shaped, c. 2 × 3cm, fleshy, endocarp with 5 valves on upper surface. Forest; 600m.
**Distr.:** S Nigeria to Gabon [lower Guinea].
**IUCN:** NT
**Bechati:** Tchiengue, 2826 9/2006.
**Uses:** FOOD — Infructescences (Tchiengue 2826).

Note: Tchiengue 2826 is a field determination (unicate at YA), name unconfirmed, Harvey, i.2010.

### Trichoscypha lucens Oliv.
Enum. Pl. Afr. Trop. 2: 230 (1992).
Treelet, to 2.5m; leaves 60cm, 11-jugate, leaflets oblong-elliptic, to 20 × 6cm, acuminate, obtuse, to 10-nerved; petiole 9cm; inflorescence terminal, glabrescent, c. 50 × 25cm, branches numerous, patent, to 13cm; flowers 2–3mm, white; stamens bright orange. Forest; 1700m.
**Distr.:** Cameroon, Gabon to Congo (Kinshasa) [lower Guinea & Congolian].
**IUCN:** LC
**Fosimondi:** Tah 294 4/2004.
Note: Kenneth Tah 294 is tentatively placed in this taxon.

## ANNONACEAE

### G. Gosline

### Monanthotaxis sp. nov. of Bechati
Climber; stems dark brown, 3mm wide, with scattered black pilose hairs, 2mm; leaves oblanceolate, white-blue below, c. 18 × 8cm, apex obtuse, base rounded-truncate then abruptly cordate, sinus 3–4mm, lateral nerves 10–12 on each side of midrib, tertiary nerves scalariform, midrib and sometimes secondaries pilose as stem; fruit with five yellow mericarps each c. 15cm, articles c. 6, each 2.5 × 0.6cm, pilose hairy. Forest; 590m.
**Distr.:** Cameroon (Bechati).
**Bechati:** Tchiengue, 2816 9/2006.

### Uvariodendron connivens (Benth.) R.E.Fr.
Tree to 13m; leaves elongate-obovate to 40 × 12cm, abruptly acuminate, glabrous, c. 20 pairs lateral veins; flowers axillary on young shoots, solitary, pedicel c. 1.5cm, sepals green c. 1cm, petals purple ovate c. 3 × 2cm tomentellous; mericarps glabrous, ellipsoid, c. 4 × 3cm. Forest understory; 1486m.
**Distr.:** SE Nigeria, Bioko & Cameroon [lower Guinea].
**IUCN:** NT
**Fosimondi:** Tchiengue, 2204 4/2005.
**Local name:** Kai (Banwa) (Tchiengue 2204).

### Xylopia africana (Benth.) Oliv.
Tree, to 12m; leaves obovate-elliptic, 9–16 × 4–8cm, strongly reticulate, nerves on underside dark-red; flowers yellow-orange, subglobose, c. 1.5cm diam., remaining closed except for a small apical opening; fruits with pedicel c. 3cm, monocarps c. 10, cylindrical, 3–12 × 1cm, pointed, constricted between seeds, stipe c. 2cm. Montane forest; 1150m.
**Distr.:** SE Nigeria, SW Cameroon & São Tomé [W Cameroon Uplands].
**IUCN:** VU
**Fosimondi:** Tchiengue, 2637 2/2006.
Note: Tchiengue 2637 comprises fallen fruit only so the identification is tentative. Better material is needed to confirm, Cheek, i.2010.

# APOCYNACEAE

M. Cheek (K), Y.B. Harvey (K), H. Beentje (K) & X. van der Burgt (K)

### *Baissea gracillima* (K.Schum.) Hua

Bull. Jard. Bot. Nat. Belg. 64: 106 (1995).

Climber, to 12m or more; exudate white to green; stem twining, tendrils absent, 2mm wide, brown with scattered patent hairs; leaves oblanceolate, to 10.5 × 4.5cm, acumen short, base obtuse to slightly cordate, lateral nerves 7–8 on each side of midrib, red when live, tertiary nerves scalariform; petiole 5mm, black scurfy; infructescence terminal, fruits pendulous, moniliform, 30–34 × 1cm, papery, with c. 10 seed swellings; seed body comma-shaped, 1.5cm, dark brown, one end with a plume of silky brown hairs 2.5cm long. Forest; 620m.

**Distr.:** Nigeria, Cameroon, Gabon, CAR, Congo (Brazzaville), Congo (Kinshasa), Equatorial Guinea, Angola [Guineo-Congolian].

**IUCN:** LC

**Bechati:** van der Burgt, 894 fr., 9/2006.

### *Pleiocarpa rostrata* Benth.

Agric. Univ. Wag. Papers 96(1): 152 (1996).

**Syn.** *Pleiocarpa talbotii* Wernham

Shrub 1–3m, glabrous; leaves opposite, lanceolate or elliptic, c. 18 × 8cm, acumen 1cm, base obtuse-decurrent, lateral nerves up to 9 pairs, petiole 1cm; fascicles 1–4-flowered, corolla white, tube 2–2.5cm, 3mm wide (pressed), anthers inserted 3–8mm from throat, lobes twisting to left; fruit apocarpous, follicles 4–5, berries, ovoid, 3cm, rounded, stipitate or not, 1–5-seeded. Forest; 1300–1350m.

**Distr.:** SE Nigeria to Gabon [lower Guinea].

**IUCN:** LC

**Fosimondi:** Tah 323 4/2004; Tchiengue, 2606 2/2006.

Note: Tchiengue 2606 is *Pleiocarpa rostrata* Benth. sensu Omino dets at K (including *P. talbotii*). However, this is a very high altitude (1300m) for this taxon. Revision not seen, Cheek, xi.2008.

### *Rauvolfia mannii* Stapf

Shrub, 1–3m, glabrous; leaves opposite or in whorls of 3, elliptic, c. 9(–15) × 3(–7)cm, acumen 1cm, cuneate-decurrent, lateral nerves 10 pairs; petiole 1cm; panicle 1.5cm, 3–10-flowered; corolla 8mm, white with 5 red spots, rarely pink; berry bilobed, rarely entire. Forest; 250–320m.

**Distr.:** tropical Africa.

**IUCN:** LC

**Bechati:** Tchiengue, 2758 9/2006; 2780a 9/2006.

### *Tabernaemontana* sp. of Bali Ngemba

The Plants of Bali Ngemba Forest Reserve: 84 (2004).

Tree, to 10m, glabrous; leaves to 23 × 9.5cm, elliptic to oblong, acute to subacuminate, cuneate, c. 15–16 pairs of lateral nerves, margin revolute; petiole to 20mm; flowers white; fruit slightly bilobed, deep longitudinal ridges, dark green with yellow spots. Disturbed forest; 1300–1610m.

**Distr.:** Cameroon (Bali Ngemba F.R., Dom & Fosimondi) [W Cameroon Uplands].

**Fosimondi:** Tah 320 4/2004; Tchiengue, 2252 4/2005; 2569 2/2006.

**Local name:** Atong (Banwa) (Tchiengue 2252).

### *Tabernaemontana* sp. of Fosimondi

Tree to 10m; white exudate; leaves 27–39 × 18–19cm, acumen 0.5cm, lateral nerves 7 on each side of the midrib; petiole 2cm; flowers with yellowish-white corolla tube. Submontane forest; 1150m.

**Distr.:** Cameroon (Fosimondi).

**Fosimondi:** Tchiengue, 2619 2/2006.

Note: only leaves are present on the Kew duplicate of Tchiengue 2619, inadequate for placing to species. Possibly *T. crassa* or *T. pachysiphon*. The leaves are much larger than those of *T.* sp. of Bali Ngemba, Cheek, i.2010.

### *Voacanga* sp. 1 of Bali Ngemba & Kupe

The Plants of Bali Ngemba Forest Reserve: 84 (2004); The Plants of Kupe, Mwanenguba and the Bakossi Mountains: 245 (2004).

Tree 6(–9)m, glabrous; stems greyish-white, white pustular lenticellate, puberulent when young; leaves oblong or oblanceolate, to 24 × 7cm, acumen 0.5cm, cuneate, lateral nerves to 14 pairs; petiole 2.5cm; flowers not seen; fruiting peduncle 8–12cm, 1-fruited, pendulous, fruit with carpels united, only slightly bilobed before dehiscence (when to 4 × 5cm), drying black, finely ridged, puberulent, dehiscing along a single terminal suture, when 4.5 × 7 × 4cm. Forest; 1700m.

**Distr.:** Cameroon (Kupe-Bakossi, Bali Ngemba & Fosimondi) [W Cameroon Uplands].

**Fosimondi:** Tchiengue, 1939 4/2004.

# ARALIACEAE

M. Cheek (K)

Fl. Cameroun 10 (1970).

### *Polyscias fulva* (Hiern) Harms

Tree, to 30m; leaves to 80cm long, imparipinnate; leaflets 3–12 paired, coriaceous, to 17 × 7.5cm, lanceolate-ovate; inflorescence of compound racemosely arranged racemes; pedicels to 5mm; petals greenish to creamy white; fruit broadly ovoid to subglobose, 3–6 × 2–5mm, ribbed. Submontane forest on slopes of hills; 1400–1800m.

**Distr.:** tropical Africa.

**IUCN:** LC

**Fosimondi:** Sight record, 2 4/2005, 2/2006.

Note: no collections made, known only from sight collections made in iv.2005 and ii.2006, Harvey, ii.2010.

*Schefflera hierniana* Harms
Epiphytic, climbing shrub 4–8m; leaves c. 7 digitately compound, c. 35cm, leaflets elliptic, 7–13 × 7cm, acumen slender, 1–2cm; persistent fleshy-leathery bracts at base of inflorescence, most conspicuous in immature inflorescences; inflorescence congested, densely brown scurfy-pubescent; partial peduncles 2–3mm; pedicels 2mm. Forest; 1400m.
**Distr.:** Cameroon & Bioko [W Cameroon Uplands].
**IUCN:** EN
**Fosimondi:** <u>Tchiengue, 1957</u> 4/2004; <u>2164</u> 4/2005.
**Uses:** ENVIRONMENTAL USES — Boundaries/Barriers/Supports (Tchiengue 2164).
Note: taxon resurrected by Frodin from synonymy with *S. barteri*; much rarer.

# ARISTOLOCHIACEAE

M. Cheek (K)

*Pararistolochia* sp. of Bechati
Climber, glabrous; leaves ovate, to 20 × 18cm, acumen 0.5cm, base cordate, sinus 4 × 4cm, nerves pedate; petiole 20cm. Forest; 320m.
**Distr.:** Cameroon (Bechati).
**Bechati:** <u>Tchiengue, 2789</u> 9/2006.

# ASCLEPIADACEAE

D. Goyder (K)

Asclepiadaceae is now generally included in the Apocynaceae, but for consistency with earlier checklists in this series, we have kept the families separate.

*Batesanthus parviflorus* Norman
J. Bot. 67, suppl. 2: 91 (1929).
Woody twiner; latex white; leaves opposite, to 11 × 6cm, oblong, glabrous, with interpetiolar stipular fringe; flowers in lax axillary panicles to 30cm long; corolla purple, rotate, lobes to 5mm, rounded; corona annular, lobed. Forest edge; 1800m.
**Distr.:** Sierra Leone to Cameroon [upper & lower Guinea].
**IUCN:** NT
**Fosimondi:** <u>Tchiengue, 1927</u> 4/2004.
Note: taxon not recognised by FWTA or Lebrun & Stork but maintained by Venter (BLFU) who is revising the genus. Species needs delimitation before a full conservation assessment.

*Kanahia laniflora* (Forssk.) R.Br.
Erect shrub, 0.5–2m; exudate white; leaves opposite, linear, 5.5–16 × 0.5–0.8cm, glabrous, apex acute, base attenuate, margin ± revolute; petiole c. 0.5cm; inflorescence one per node between leaf bases; peduncle 3–4.5cm, crowned by numerous bracts; bracts to 1cm, narrowly triangular; flowers white, 5–15 in succession; pedicels 1.5–2cm; sepals 1.5–4.5mm.

narrowly triangular; corolla 5 lobed; lobes patent, woolly hairy, elliptic-oblong, 5–15mm; corona 3–5.5mm high; stipe 0–2.5mm high; style-cap level with corona lobes or raised above fruit; fruit with follicles mostly paired, fusiform, c. 5 × 1cm. An obligate rheophyte occuring here in rocky mountain streams; 250m.
**Distr.:** Ivory Coast to Somalia, N Sudan to South Africa, Arabia [tropical Africa].
**IUCN:** LC
**Bechati:** <u>Tchiengue, 2770</u> 9/2006.

*Marsdenia angolensis* N.E.Br.
**Syn.** *Gongronema angolense* (N.E.Br.) Bullock
Herbaceous or woody scrambler, to 4m; latex white; leaves 5–12 × 3–7cm, broadly ovate, acuminate, deeply cordate, pubescent; petiole 2–6cm long; inflorescences extra-axillary, about as long as the petioles of the adjacent leaf pair, with 2–3 principal branches terminated by subumbelliform clusters of flowers, pubescent; pedicels 4–10mm; sepals c. 1.5–2mm long; corolla cream or yellowish green, united into a broadly cylindrical tube for about half its length; tube 1.5–2.5mm long; lobes 1.5–2 × 1.5mm; corona lobes c. 2mm, fleshy; follicles occurring singly or paired, narrowly cylindrical and tapering to a slightly displaced tip, 8–12 × 0.5cm, densely pubescent with short spreading hairs; seeds flattened, c. 7 × 2mm, oblong with a narrow yellowish margin. Forest understorey.
**Distr.:** tropical Africa.
**IUCN:** LC
**Fosimondi:** <u>Tchiengue, 2176</u> 4/2005.

*Marsdenia latifolia* (Benth.) K. Schum.
Omlor, Gen. Rev. Marsdenieae (Asclepiadaceae): 75 (1998).
**Syn.** *Gongronema latifolium* Benth.
Slender to robust woody scrambler; latex white; leaves opposite, to c. 12 × 7cm, ovate to ovate-oblong, pubescent; inflorescences extra-axillary with cream or yellow-green flowers scattered along the axes; corolla lobes c. 2–2.5mm long; corona lobes fleshy and with an apical tongue; follicles single, subcylindrical. Margins of evergreen forest or scrub; 350m.
**Distr.:** widespread in C and W Africa, to Zambia, Angola [tropical Africa].
**IUCN:** LC
**Bechati:** <u>Tchiengue, 2764</u> 9/2006.

# BALSAMINACEAE

J. Arcate, D. Haynes, K. Herian, M. Corcoran (K), M. Cheek (K), H. Sook Choung & B. Tchiengué (YA)

Fl. Cameroun 22 (1981).

*Impatiens burtoni* Hook.f.
Terrestrial herb, c. 30cm; leaves alternate, lanceolate, c.

8 × 5cm; single-flowered; pedicel c. 3cm; flower white, c. 3 × 1cm, face concave; sepals densely long-hairy; spur about as long as petals. Farmbush; 1800m.
**Distr.:** Cameroon to Burundi [lower Guinea & Congolian].
**IUCN:** LC
**Fosimondi:** Tchiengue, 2282 4/2005.
Note: common in farmbush, notable for having large white hairy flowers.

### *Impatiens filicornu* Hook.f.
Terrestrial or epilithic, rarely epiphytic, herb c. 30cm; leaves drying dark-brown above, alternate, lanceolate, c. 9 × 4cm; peduncle c. 11cm; flowers fascicled at apex; bracts ovate, 3–4mm; pedicel 1.5cm; flowers pink to white with purple markings, face flat, c. 1.5cm diam., spur c. 15mm. Forest; 1700–1800m.
**Distr.:** Bioko, Cameroon to Cabinda [lower Guinea].
**IUCN:** LC
**Fosimondi:** Tah 288 4/2004; Tchiengue, 2250 4/2005.

### *Impatiens kamerunensis* Warb. subsp. *kamerunensis*
Grey-Wilson C., Impatiens of Africa: 164 (1980); Fl. Cameroun 22: 9 (1981).
Terrestrial or epilithic herb, c. 30cm resembling *I. filicornu*, but leaves opposite, drying green above; flowers pink, spread along the upper part of the peduncle; bracts filiform. Wet rocks by streams in forest; 320–1800m.
**Distr.:** Ghana to Cameroon [upper & lower Guinea].
**IUCN:** LC
**Bechati:** Tchiengue, 2775 9/2006; van der Waarde, 9 10/2004; **Fosimondi:** Tchiengue, 1959 4/2004.
Note: Tchiengue 2775 has opposite leaves, Cheek, xi.2008.

### *Impatiens letouzeyi* Grey-Wilson
Flore Du Cameroon 22: 25 (1981).
Epiphytic herb 50–150cm; leaves alternate, oblong-elliptic, 14 × 5cm; peduncles 0.4cm, 1–4-flowered, pedicels 4cm; flowers pink, face concave, c. 6 × 4cm, lateral spur 1cm, dentate; spur c. 6mm, tightly coiled. Forest; 1350m.
**Distr.:** Cameroon [W Cameroon Uplands].
**IUCN:** EN
**Fosimondi:** Tchiengue, 2236 4/2005.
Note: This is a big range extension from Bakossi! But Kew has a specimen from Nkom-Wum — perhaps a different subsp.? Cheek, xi.2008.

### *Impatiens mackeyana* Hook.f. subsp. *zenkeri* (Warb.) Grey-Wilson
Grey-Wilson C., Impatiens of Africa: 221 (1980).
**Syn.** *Impatiens zenkeri* Warb.
Erect terrestrial perennial, to 1m; stem simple or moderately branched, becoming glabrescent with age; leaves broadly ovate to broadly elliptic-oblong, 5.0–18.5 × 2.5–8.5cm, margin serrate; petiole without

stipitate glands; flowers solitary, pedunculate; mauvish-pink or purplish; lateral united petals considerably longer than the lower sepal; ovary glabrous; fruit 17–20 × 5–6.5mm, fusiform, glabrous. Along river margins and moist sites; 800m.
**Distr.:** Cameroon, Congo (Kinshasa), Gabon [lower Guinea].
**IUCN:** LC
**Bechati:** van der Waarde, 8 10/2004.

### *Impatiens mannii* Hook.f.
**Syn.** *Impatiens deistelii* Gilg
Terrestrial herb 0.5–1m, glabrous; leaves alternate, ovate, c. 7 × 5cm; single-flowered, pedicel c. 1cm; flower white, the lower sepal with conspicuous transverse purple bars, face concave, 1.5–2 × 0.6cm, glabrous, spur shorter than petals. Open areas in forest; 1580m.
**Distr.:** Bioko & Cameroon to Congo (Kinshasa) [lower Guinea & Congolian].
**IUCN:** LC
**Fosimondi:** Tchiengue, 2199 4/2005.

### *Impatiens sakeriana* Hook.f.
Terrestrial herb, 0.3–1m; leaves in whorls of 3–4 or opposite, ovate-elliptic, c. 7.5–12 × 3.5–4cm, apex acuminate, margin crenate, pale brown-hairy below; racemes short, 3–4-flowered; peduncle c. 4cm; flowers 1.5–3cm long, lip red, gradually narrowing into short spur, greenish-red, swollen at tip. Forest patches; 1700m.
**Distr.:** Bioko & Cameroon [W Cameroon Uplands].
**IUCN:** VU
**Fosimondi:** Tah 290 4/2004.

## BASELLACEAE

M. Cheek (K)

### *Basella alba* L.
Climbing herb, to 4m, glabrous; stem twining; leaves alternate, ovate, c. 6 × 4.5cm, gradually acuminate, base truncate to cordate; petiole 3cm; stipules absent; inflorescence axillary, 8cm, several-flowered; flowers white and purple, scarcely opening, 4mm. Farm in submontane forest; 1486m.
**Distr.:** Sierra Leone to Kenya [tropical Africa].
**IUCN:** LC
**Fosimondi:** Tchiengue, 2203 4/2005.

## BEGONIACEAE

M. Sosef (WAG)

### *Begonia adpressa* Sosef
Studies in Begoniaceae 5. Wag. Agric. Univ. Papers (1994).
Terrestrial herb, 20cm; blade ovate, c. 12 × 7cm, acuminate, entire; petiole c. 12cm adpressed-hairy;

peduncle 15cm, 2-flowered; perianth 1.5cm, yellow. Forest; 1290m.
**Distr.:** Cameroon [W Cameroon Uplands].
**IUCN:** VU
**Fosimondi:** Tchiengue, 2638 2/2006.
Note: Tchiengue 2638 is tentatively placed in this taxon, owing to the seemingly patent pubescence on the veins, Sosef, i.2009.

## *Begonia ampla* Hook.f.
Epiphytic herb, 1–5m from ground, stems 1cm diam. or more, scales fimbriate, silvery; blades pale green, suborbicular, 22cm, acuminate, cordate; petiole c. 17cm; inflorescence concealed in cup-like bracts 3–7cm wide; perianth 1.5cm, white striped pink; fruit berry-like, globose, indehiscent, c. 1cm diam. Forest; 1400m.
**Distr.:** Cameroon to Uganda [afromontane].
**IUCN:** LC
**Fosimondi:** Tah 313 4/2004.

## *Begonia eminii* Warb.
Stoloniferous hemiepiphyte; leaves oblong-elliptic, c. 11 × 5cm, base cordate, subequal; female inflorescence 2-flowered; flowers pink; fruit 3-angular, red; dehiscing by slits or valves. Forest; 1300–1350m.
**Distr.:** Cameroon to Uganda [lower Guinea & Congolian].
**IUCN:** LC
**Fosimondi:** Tah 301 4/2004; Tchiengue, 2615 2/2006.

## *Begonia oxyanthera* Warb.
**Syn.** *Begonia jussiaeicarpa* Warb.
Stoloniferous epiphyte; leaves elliptic, c. 10 × 4cm, acuminate, obtuse to rounded, entire or slightly dentate, lower margin and nerves red, lower surface glabrescent; petiole 1cm; inflorescence 1.5cm; perianth 0.8cm, cream-red; fruit cylindric, 4 × 0.4cm, red. Forest; 1460–1800m.
**Distr.:** Bioko & Cameroon [W Cameroon Uplands].
**IUCN:** VU
**Fosimondi:** Tchiengue, 1911 4/2004; 2276 4/2005.

## *Begonia oxyloba* Welw. ex Hook.f.
Terrestrial herb, 0.3–1m; stem ascending; leaves palmately 5–7-lobed to about half the blade radius, transversely unequally elliptic in outline, c. 19 × 23cm, base shallowly cordate; petiole c. 14cm; inflorescence axillary, 3–4-flowered, 3cm; flowers pink, c. 2cm; fruit ellipsoid, juicy, indehiscent, 2cm. Forest; 1700–1800m.
**Distr.:** Guinea (Conakry) to Tanzania & Madagascar [tropical Africa & Madagascar].
**IUCN:** LC
**Fosimondi:** Tah 295 4/2004; Tchiengue, 1962 4/2004; 2248 4/2005.

## *Begonia poculifera* Hook.f. var. *poculifera*
Epiphyte, resembling *B. ampla*, but blade axis at right angles to petiole, dark green veined red, obliquely ovate-lanceolate, 11 × 5.5cm, acuminate, truncate, entire. Forest; 1800m.

**Distr.:** SE Nigeria to Tanzania [afromontane].
**IUCN:** LC
**Fosimondi:** Tchiengue, 1912 4/2004.

## *Begonia preussii* Warb.
Engl. Bot. Jahrb. 22: 36 (1895).
**Syn.** *Begonia sessilanthera* Warb.
Epiphytic herb resembling *B. oxyanthera* but occurring at lower altitudes, peduncle 0.5–5cm, 7–15-flowered (not 0.1–0.8cm, less than 7-flowered), perianth pink or red (not white, red-edged), anthers 1.5mm, truncate (not 2.7–4.5mm, acute or acuminate). Forest; 1486m.
**Distr.:** Bioko, SE Nigeria & Cameroon [W Cameroon Uplands].
**IUCN:** VU
**Fosimondi:** Tchiengue, 2188 4/2005.

## *Begonia pseudoviola* Gilg
Terrestrial herb, 14cm, lacking aerial stems; blade transversely ovate, 6cm, acute, cordate, subserrate, upper surface usually purple-black, with long white patent hairs; petiole 10cm; peduncle 3.5cm; perianth 1cm, yellow. Forest; 250–400m.
**Distr.:** Cameroon [W Cameroon Uplands].
**IUCN:** VU
**Bechati:** van der Burgt, 869 fl., 9/2006; Tchiengue, 2753 fl., 9/2006; 2797 9/2006; van der Waarde, 2 10/2004.

## *Begonia quadrialata* Warb. subsp. *quadrialata* var. *quadrialata*
Terrestrial herb, 20cm; lacking aerial stems; blade suborbicular, 9 × 7.5cm, obtuse, entire; petiole c. 10cm, hairy; peduncle c. 6cm; perianth 0.5cm, yellow. Forest.
**Distr.:** Sierra Leone to Togo, SE Nigeria to Angola [upper & lower Guinea].
**IUCN:** LC
**Fosimondi:** Tchiengue, 2247 4/2005.

## *Begonia schaeferi* Engl.
Engl. Pflanzenw. Afr. iii. II. (Engl. & Drude, Veg. der Erde, ix.) 618 (1921).
Terrestrial herb, 10–20cm; lacking an aerial stem; rhizome creeping; leaves peltate, ovate, 3.5–12 × 2.5–6cm, apex long acuminate, margin and upper surface sparsely hairy with straight hairs; petioles 4–18cm; inflorescence of two sorts, one male, the other bisexual, branched or not, nearly as long as leaves, apex with one or more pairs of patent persistent elliptic bracts 4–8mm, margin ciliate; flowers yellow, the segments 7–14mm long. Submontane forest; 1540m.
**Distr.:** SE Nigeria & Cameroon [W Cameroon Uplands].
**IUCN:** VU
**Fosimondi:** Tchiengue, 2576 2/2006.
Note: Tchiengue 2576 is tentatively placed in this taxon, the bracts are too large for *B. quadrialata*, Sosef, i.2009.

**Begonia scutifolia** Hook.f.
Agric. Univ. Wag. Papers 94(1): 206 (1994).
Terrestrial herb, to 10cm; lacking an aerial stem;
rhizome creeping; leaves peltate; asymmetric, ovate less
usually sublinear, elliptic, 5–10 × 0.5–5cm, apex acutely
tapering, entire upper surface glabrous apart from
minute sausage-like hairs; petiole up to 1.5cm long,
sparsely to densely hirsute with patent or adpressed
hairs; inflorescence with 1–3 male and 1 female flowers,
2–7.5cm; peduncle unbranched; petals yellow, the inner
with a basal red patch inside, 5–10mm; ovary wings
narrow, ribbon-like, inconspicuous; fruit/ovary
length:breadth c. 5:1. Forest, sometimes on rocks.
**Distr.:** Cameroon, Gabon, Congo (Kinshasa) and
Cabinda [lower Guinea].
**IUCN:** LC
**Fosimondi:** Tchiengue, 2229 4/2005.
Note: Tchiengue 2229 is tentatively placed in this
taxon, specimen is sterile and flowers are needed,
Sosef, i.2009.

## BUDDLEJACEAE

Y.B. Harvey (K)

**Nuxia congesta** R.Br. ex Fresen.
**Syn.** *Lachnopylis mannii* (Gilg) Hutch. & M.B.Moss
Tree, 2–8m; bole white, fibrous; stems orange-brown;
leaves in whorls of 3, dimorphic: type 1 ovate-elliptic,
c. 4 × 2.5cm, apex rounded, base cuneate, serrate,
petiole 0.2cm; type 2 c. 11 × 5.5cm, entire, petiole 2cm;
inflorescence terminal, 7 × 10cm, dense, many-
flowered; flowers white, c. 7mm. Forest & forest edge;
2100–2150m.
**Distr.:** Guinea (Conakry) to South Africa [afromontane].
**IUCN:** LC
**Fosimondi:** Sight record, 3 11/1974.
Note: no collections made, but taxon mentioned in the
habitat notes of Letouzey 13382 & 13386, Harvey,
ii.2010.

## BURSERACEAE

Y.B. Harvey (K) & B. Tchiengué (YA)

**Canarium schweinfurthii** Engl.
Tree, to 40m; copious exudate, drying whitish; young
branches and young leaves rusty-pubescent; leaves
clustered towards branch apices, c. 30–60cm long,
imparipinnate, 8–12-jugate; leaflets oblong-ovate, 5–20
× 3–5.5cm, base cordate, lateral nerves pubescent
below; petiole flat on upper side towards base; panicles
to 30cm long; pedicels c. 3mm; flowers cream-white, c.
1cm; fruits ellipsoid, c. 3.5 × 2cm, epicarp glabrous,
purple-black, endocarp trigonous. Forest; 470–560m.
**Distr.:** Senegal to Sudan & Tanzania [Guineo-Congolian].
**IUCN:** LC
**Bechati:** Tchiengue, 2771 9/2006; **Fosimondi:** Sight
record, 4 9/2006.

Note: Tchiengue 2771 is a unicate and the
determination has not been confirmed at Kew, in
addition, Tchiengue records the presence of this taxa in
the habitat notes of No. 2809, Harvey, ii.2010.

**Dacryodes edulis** (G.Don) H.J.Lam
Tree to 20m; exudate brown, branchlets, petioles,
inflorescence rachis and outside of flowers stellately
brown-hairy; leaves imparipinnate, 4–7-jugate, leaflets
oblong-ovate, 12.5–20.5 × 3.5–5.5cm, acumen to
1.8cm, base acute, oblique in lateral leaflets,
undersurface sparsely stellate-hairy with long simple
hairs along midrib, lateral nerves 11(–14) pairs; petiole
9cm; petiolules 0.5cm; panicles many-branched, partial
peduncles to 14cm, grooved; flowers 0.4cm, cream;
fruit ovoid-ellipsoid, 5.5 × 2.2cm. Forest, and planted
near villages; 1350m.
**Distr.:** S Nigeria to Angola [lower Guinea & Congolian].
**IUCN:** LC
**Fosimondi:** Tah 315 4/2004.
**Local name:** African plum (Tah 315).

**Santiria trimera** (Oliv.) Aubrév.
Tree, to 20m; laterally-flattened stilt roots, glabrous;
stem strongly smelling of turpentine when cut; leaves
imparipinnate, 2–3-jugate, leaflets oblong-elliptic,
10.5–16.5 × 4–6.5cm, acumen narrow, to 2.2cm long,
base acute to obtuse, lateral nerves 7–10 pairs; petiole
9.5cm; petiolules 1cm; panicles axillary, to 9cm; fruit
green-black, asymmetrically oblate, 1.2 × 2 × 1.5cm,
style-remains lateral; fruiting pedicel 0.8cm. Forest;
1380m.
**Distr.:** Sierra Leone to Congo (Kinshasa) [Guineo-
Congolian].
**IUCN:** LC
**Fosimondi:** Tchiengue, 2214 4/2005.
**Local name:** Lai (Banwa) (Tchiengue 2214).
Note: see Onana, J.-M. (2009). Le genre *Santiria*
(Burseraceae) en Afrique: Redéfinition de *Santiria*
*trimera*. Syst. Geogr. Pl. 79: 215–224.

## CAMPANULACEAE

Y.B. Harvey (K) & B. Tchiengué (YA)

**Lobelia columnaris** Hook.f.
Herb, 2m tall, unbranched; stem 2cm diam. at base,
glabrous; leaves oblanceolate-oblong, c. 30 × 8cm at
base of stem, gradually diminishing in size towards the
apical inflorescence, apex acute, margin
inconspicuously serrate, softly hairy; inflorescence a
densely-flowered, unbranched spike, occupying the
apical half of the plant; corolla pale blue, lilac to white,
2–3cm long. Forest edges and grassland; 1700m.
**Distr.:** Bioko, Cameroon and Nigeria [W Cameroon
Uplands (montane)].
**IUCN:** NT
**Fosimondi:** Tchiengue, 1938 4/2004.

### Lobelia hartlaubii Buchenau

Straggling or erect herb 15–35(–50)cm long; stems ribbed, subglabrous, often rooting at the lower nodes; leaves triangular, 15–35 × 10–30mm, margin coarsely dentate; petiole 5–20mm long, occasionally "winged"; racemes 2–4-flowered; pedicels c. 4mm long; bracts 3–8mm long; bracteoles c. 1.5mm long, near top of pedicel; hypanthium to 4mm long in flower, extending to 9mm in fruit; calyx lobes linear; corolla c. 6mm long; seeds 0.7–1mm long, numerous, pale or reddish brown. Grassland-forest edge, forest; 1386m.
**Distr.:** Nigeria to Uganda & Burundi, Madagascar [tropical Africa].
**IUCN:** LC
**Fosimondi:** Tchiengue, 2195 4/2005.

### Lobelia rubescens De Wild.

**Syn.** *Lobelia kamerunensis* Engl. ex Hutch. & Dalziel
Straggling herb, to 45cm; stems triangular with narrowly-winged edges, often rooting at nodes; leaves lanceolate, gradually narrowing upwards on stem, 13–30 × 3–10mm, margin serrate, subglabrous; petioles c. 5mm long; flowers in leafy racemes; pedicels 7–14mm long; bracts to 14mm long; hypanthium 5–6mm long (extending to 7mm in fruit); calyx lobes narrowly triangular; corolla to 10mm long, blue with yellow throat; seeds c. 0.5mm long, brown. Farmbush, grassland; 1800m.
**Distr.:** Nigeria to Tanzania [lower Guinea & Congolian].
**IUCN:** LC
**Fosimondi:** Tchiengue, 1968 4/2004.

## CECROPIACEAE

M. Cheek (K)

Fl. Cameroun 28 (1985).

### Myrianthus arboreus P.Beauv.

Tree, 7(–25)m; leaves alternate, palmately compound; leaflets 7, elliptic oblong c. 30 × 10cm, acute or acuminate, serrate, concolorous; petiole robust, 30cm; male inflorescence c. 20 × 15cm; fruit compound, ellipsoid, c. 15 × 10cm, yellow, juicy. Forest, farmbush; 620–670m.
**Distr.:** Guinea (Conakry) to W Tanzania [Guineo-Congolian].
**IUCN:** LC
**Bechati:** van der Burgt, 899 9/2006; Fosimondi: Sight record, 5 9/2006.

### Myrianthus fosi Cheek

The Plants of Lebialem Highlands, Cameroon: 59 (2010)
**Syn.** *Myrianthus* sp. 1 of Kupe
The plants of Kupe, Mwanenguba and the Bakossi Mountains, Cameroon: 260 (2004).
Treelet, 3–5m; leaves alternate, palmately 3–5-lobed by × the radius, white below with brown-hairy nerves;

petioles unequal in length on same stem. Submontane forest; 1290m.
**Distr.:** Cameroon [Cameroon endemic].
**IUCN:** VU
**Fosimondi:** Tchiengue, 2623 2/2006.

### Myrianthus preussii Engl. subsp. *preussii*

Treelet, 2–4(–9)m; leaves alternate, palmately compound; leaflets 5(–7), bright-white below, obovate, long-acuminate; petioles slender, 15cm; petiolules c. 3cm; male inflorescence c. 6 × 4cm; fruit globose, c. 5cm diam. Forest; 590m.
**Distr.:** SE Nigeria & Cameroon [lower Guinea].
**IUCN:** NT
**Bechati:** Tchiengue, 2838 9/2006.

## CELASTRACEAE

G. Gosline, B. Tchiengué (YA) & X. van der Burgt (K)

Fl. Cameroun 19 (1975) & 32 (1990).

### Salacia erecta (G.Don) Walp. var. *erecta*

Climber, to 15m, no resinous threads; branchlets dark brown smooth; leaf blade elliptic, 4–15 × 1.5–5cm, 7–10 secondary nerves, margin toothed; petioles to 5mm, channel margins undulate; flowers in sessile axillary fascicles, sometimes on thin shoots appearing like peduncles; pedicels 3–7mm; buds ovoid; flowers c. 5mm diam., yellow; fruits globular, 1–3cm diam., orange-red. Forest; 1800m.
**Distr.:** Guinea (Conakry) to Zambia [tropical Africa].
**IUCN:** LC
**Fosimondi:** Tchiengue, 1916 4/2004.

### Salacia lebrunii Wilczek

Fl. Gabon 29: 133 (1986).
Climber, 15m+; stems 4-ridged, green; leaves opposite, blades elliptic, 7–9.5 × 3.2–5cm, acumen 0.5cm, base obtuse, abruptly rounded, lateral nerves 3–6 pairs, teeth inconspicuous; petiole 3mm, dilated; flowers axillary among leaves, single; pedicel 8mm; petals 5, light orange, 2 × 1mm; stamens 2 (usually 3 in Salacia) folded against style, later curving in a half circle downwards, orange; ovary green to orange; style conical; fruit subglobse, hard, orange, 28 × 26 × 23mm; seeds 2, large. Forest, along stream; 400m.
**Distr.:** Cameroon, Gabon, Congo (Kinshasa) [Guineo-Congolian].
**IUCN:** EN
**Bechati:** van der Burgt, 870 fl., fr., 9/2006.

### Salacia lehmbachii Loes. var. *pes-ranulae* N.Hallé

Fl. Gabon 29: 75 (1986); Fl. Cameroun 32: 55 (1990).
Shrub, to 2m; without resinous threads, stems quadrangular; leaves sessile, ellongate-elliptic to 20 × 7cm, linear acumen 2.5cm long, c. 6 pairs secondary veins, tertiary veins perpendicular to midrib across leaf;

infl. sometimes in midrib grooves, with alternate divisions starting 4–20mm below summit of rachis-peduncle; flowers 5–8mm diam., red to orange; fruit long and acuminate. Forest; 630m.
**Distr.:** SE Nigeria & Cameroon [lower Guinea].
**IUCN:** VU
**Bechati:** Tchiengue, 2808 9/2006.

*Salacia pynaertii* De Wild.
Ann. Mus. Congo (ser.) 5(ii): 295 (1908); Flora Zambesiaca (online) vol. 2(2): 402 (1966).
Liane, to 6m, glabrous; stems flattened; leaves opposite to subopposite; blade 6–13 × 3–6cm, oblong. elliptic-oblong to ovate; petiole 4–7mm long; flowers c. 8–18, sessile or shortly pedunculate fascicles; pedicels 3–5mm long; petals 1.5–2mm long; fruit orange-pink with small white spots, 1–2cm diam., smooth, 1–3-seeded. Evergreen forest and forest edges.
**Distr.:** Cameroon, Congo (Kinshasa), Nigeria, CAR, Gabon, Angola & Zambia [Guineo-Congolian].
**IUCN:** LC
**Fosimondi:** Tchiengue, 2288 4/2005.
Note: Tchiengue 2288 is tentatively placed in this taxon, Tchiengue, x.2008.

*Salacia* sp. aff. *nitida* of Fosimondi (Benth.) N.E.Br.
Climber, glabrous; stems finely ridged when dry, internodes 5cm; leaf-blades elliptic, c. 14 × 5cm, acumen 1cm, base acute-attenuate, lateral nerves c. 6 pairs, tertiary nerves scalariform, serrate-dentate; petiole 2–3mm; inflorescence in axils of leaves; peduncle 10mm; apical bract cluster 1mm; pedicels 8mm; flowers orange, 5mm diam. Submontane forest; 1800m.
**Distr.:** Cameroon (Fosimondi) [Cameroon Endemic].
**Fosimondi:** Tchiengue, 1928 4/2004.
Note: known only from Fosimondi, more material needed, Cheek, i.2010.

# CHRYSOBALANACEAE

M. Cheek (K)

Fl. Cameroun 20 (1978).

*Magnistipula conrauana* Engl.
Fl. Cameroun 20: 79 (1978).
**Syn.** *Hirtella conrauana* (Engl.) A.Chev.
Shrub or tree, to 12m; branches glabrous, lenticellate; stipules foliaceous, ovate, base oblique, 3–4 × 2–3cm; leaves ovate to oblong-elliptic, 22–25 × 10–12cm, base acute to rounded, decurrent, glabrous, numerous glands towards base and under acumen, lateral veins 6–8 pairs; panicles spreading, terminal and axillary, to 30cm, often subtended by foliaceous bracts, rachis glabrous; normal bracts 3 × 2.5mm, triangular, margins glandular; bracteoles 1mm, eglandular; pedicels 4–5mm, receptacle obliquely campanulate, curved, 6–7mm, glabrous outside, deflexed-villous within; sepals 4 × 2.5mm, ciliate, tomentellous within; petals white to pale

violet, 2–4mm; stamens 7, 4–6mm long; staminodes c. 7, tooth-like; ovary glabrous; style curved, 3–5mm; fruit ovoid, 20 × 35 × 55mm. Forest; 1800m.
**Distr.:** Cameroon [lower Guinea].
**IUCN:** EN
**Fosimondi:** Tchiengue, 1934 4/2004.

# COMPOSITAE

D.J.N. Hind (K), B. Tchiengué (YA) & H. Beentje (K)

*Acmella caulirhiza* Delile
Fragm. Flor. Geobot. 36(1) Suppl.1: 228 (1991); F.T.E.A. Compositae 3: 730 (2005).
**Syn.** *Spilanthes filicaulis* (Schum. & Thonn.) C.D.Adams
Annual or perennial creeping herb, 5–30cm tall; leaves ovate, up to 5 × 2.5cm, obtuse, subentire to dentate; petiole 0.25–1cm; peduncles to 10cm; capitula conical; ray florets present, sometimes tiny, yellow; pappus usually absent. Forest & stream banks; 1400m.
**Distr.:** tropical & subtropical Africa and Madagascar [tropical & subtropical Africa].
**IUCN:** LC
**Fosimondi:** Tchiengue, 2600 2/2006.

*Bidens barteri* (Oliv. & Hiern) T.G.J.Rayner
Kew Bull. 48: 483 (1993).
**Syn.** *Coreopsis barteri* Oliv. & Hiern
Annual herb, 0.3–1m tall; stems purple; leaves lanceolate, c. 5 × 1.5cm, apex acute, base rounded, margin serrate, glabrous; petiole 1–2mm long; capitula few, 3cm diam., radiate; rays yellow, clawed, 14 × 4mm; disc florets black; involucre of two whorls of c. 10 phyllaries each, inner broader, black, outer green. Farmbush; 1800m.
**Distr.:** Ghana to Cameroon [lower Guinea].
**IUCN:** LC
**Fosimondi:** van der Waarde, 16 10/2004.
Note: van der Waarde 16 is tentatively placed in this taxon, Hind, ii.2010.

*Crassocephalum bougheyanum* C.D.Adams
Herb, 1–2m tall; cauline leaves simple, ovate, c. 6 × 4cm (lowest leaves sometimes trilobed), apex acute, base truncate to obtuse, margin bidentate; petiole 1–2cm long; capitula discoid, 1–3cm diam., corollas orange or red; pappus of white bristles. Farmbush or forest edge; 2100m.
**Distr.:** Mt Cameroon, Bamenda Highlands and Bioko [W Cameroon Uplands (montane)].
**IUCN:** NT
**Fosimondi:** Letouzey 13386 11/1974.

*Crassocephalum montuosum* (S.Moore) Milne-Redh.
Herb, 1.5m tall; cauline leaves elliptic in outline, 8.5 × 6cm, pinnately 2–8-lobed, apical lobe largest, lower

lobes decreasing in size, lobes dentate, glabrous; capitula 10–20, in aggregations, discoid, 2–4mm diam.; florets yellow or brownish yellow. Forest & forest-grassland transition; 1486m.
**Distr.:** W Cameroon to Congo (Kinshasa) to E Africa & Madagascar [afromontane].
**IUCN:** LC
**Fosimondi:** Tchiengue, 2191 4/2005.

***Emilia coccinea*** (Sims) G.Don
Annual herb, 15–120cm high; leaves ± subsucculent, 1–20 × 0.4–6cm, often purplish beneath; capitula in terminal corymbs; discoid, phyllaries uniseriate, corollas bright orange; pappus of white bristles. Forest & farmbush; 1120m.
**Distr.:** tropical Africa [afromontane].
**IUCN:** LC
**Fosimondi:** Tchiengue, 2634 2/2006.

***Helichrysum cameroonense*** Hutch. & Dalziel
Aromatic herb, to 1.2m; leaves glandular-hairy; capitula many, dense, disciform, to 2.5cm diam. with pale yellow bracts and small yellow florets; pappus of white bristles. Grassland, forest-grassland transition; 2150m.
**Distr.:** W Cameroon [Western Cameroon Uplands (montane)].
**IUCN:** NT
**Fosimondi:** Letouzey 13382 11/1974.
Note: detailed information on this rare species is given in Cheek *et al.* 2000: 59.

***Lactuca inermis*** Forssk.
Kew Bull. 39: 132 (1984).
**Syn.** *Lactuca capensis* Thunb.
Perennial herb, to 2m; with white sap; leaves glaucous, lanceolate, often toothed, to 20 × 1.6cm; capitula in leafy panicles, with small blue or mauve ligulate flowers; pappus of white bristles. Grassland; 2310m.
**Distr.:** tropical & subtropical Africa [afromontane].
**IUCN:** LC
**Fosimondi:** Tchiengue, 2175 4/2005.

***Laggera crispata*** (Vahl) Hepper & J.R.I.Wood
Kew Bull. 38: 83 (1983); Fl. Masc. Compositae: 60 (1993); F.T.E.A. Compositae 2: 353 (2002).
**Syn.** *Blumea crispata* (Vahl) Merxm. var. *crispata* Fl. Afr. Cent. Compositae: 13 (1989).
**Syn.** *Laggera alata* (D.Don) Sch. Bip. ex Oliv. var. *alata*
**Syn.** *Laggera pterodonta* (DC.) Sch.Bip. ex Oliv.
Annual to perennial aromatic herb, to 2.4m; stems winged; leaves often tinged purple, obovate, to 20 × 8cm, with serrate margins; capitula in panicles, with many small pink to purple tubular flowers; pappus of white bristles. Forest; 1400m.
**Distr.:** tropical Africa & Asia [palaeotropical].
**IUCN:** LC
**Fosimondi:** Tchiengue, 2642 2/2006.

***Mikania chenopodifolia*** Willd.
Fragm. Flor. Geobot. 36(1) Suppl.1: 460 (1991); F.T.E.A. Compositae 3: 835 (2005).
**Syn.** *Mikania capensis* DC. Fragm. Flor. Geobot. 36(1) Suppl.1: 458 (1991).
Shrub 5–9m; leaves opposite, ovate, 2–10 × 1–7cm, subsaggittate to hastate, glandular-punctate, with 3 main veins; capitula in a dense, leafy corymb; phyllaries 4; florets 4, corollas white, tubular; pappus of white bristles. Forest & thicket; 670m.
**Distr.:** tropical Africa & Asia [palaeotropical].
**IUCN:** LC
**Bechati:** van der Burgt, 896 fl., 9/2006.
Note: mentioned in FWTA 2: 286 (1963) as a synonym of *M. cordata* (Burm. f.) B.L.Robinson var. *cordata*.

***Senecio purpureus*** L.
Kew Bull. 41: 904 (1986); F.T.E.A. Compositae 3: 668 (2005).
**Syn.** *Senecio clarenceanus* Hook.f.
Perennial herb, to 2.4m; leaves oblong or lyrate, 10–60 × 2–10cm; capitula in corymbs, small, discoid, corollas red or purple; ray flowers absent; pappus of white bristles. Grassland and forest-grassland transition.
**Distr.:** Mount Cameroon, Bioko, Congo (Kinshasa), Angola, Tanzania and southern Africa [afromontane].
**IUCN:** LC
**Fosimondi:** Tchiengue, 2165 4/2005.

***Vernonia hymenolepis*** A.Rich.
Kew Bull. 43: 237 (1988); F.T.E.A. Compositae 2: 240 (2002).
**Syn.** *Vernonia leucocalyx* O.Hoffm. var. *acuta* C.D.Adams
**Syn.** *Vernonia leucocalyx* O.Hoffm. var. *leucocalyx*
**Syn.** *Vernonia insignis* (Hook.f.) Oliv. & Hiern
Woody herb or shrub, to 4m; leaves lanceolate, serrate, 5–20 × 1–7cm, pubescent beneath; capitula large in large terminal corymbs; phyllaries with white or pink appendages; corollas small, mauve or purple; pappus of white bristles. Forest-grassland transition; 1640m.
**Distr.:** Cameroon, Sudan, Ethiopia, Uganda & Kenya [afromontane].
**IUCN:** LC
**Fosimondi:** Tchiengue, 2244 4/2005; 2245 4/2005.

# CONNARACEAE

L. Pearce (K) & B. Tchiengué (YA)

See "The Connaraceae: a taxonomic study with the emphasis on Africa" in Agric. Univ. Wag. Papers 89(6): 1–403 (1989) for full species account and synonymy.

***Cnestis corniculata*** Lam.
**Syn.** *Cnestis aurantiaca* Gilg
**Syn.** *Cnestis congolana* De Wild.
**Syn.** *Cnestis grisea* Baker
**Syn.** *Cnestis longiflora* Schellenb.

**Syn. *Cnestis* sp. A *sensu* Hepper**
Liana, to 10m; branchlets yellow-brown pilose to
glabrous; leaves imparipinnate, to 45cm, 7–12-jugate,
leaflets opposite to alternate, oblong, to 11 × 4.5cm,
acumen 1cm, base asymmetric, truncate or subcordate,
veins beneath pilose, basal leaflets much reduced,
ovate; petiolules 1mm; racemes axillary or cauliflorous,
to 10cm, single or in clusters of up to 4, pubescent;
flowers 0.5cm, sepals reflexed, puberulent; petals
oblong-lanceolate; follicles obliquely ellipsoid with a
curved beak at maturity, to 1 × 3cm, golden brown-
pilose. Forest & thickets; 540m.
**Distr.:** Senegal to E Tanzania [tropical Africa].
**IUCN:** LC
**Bechati:** Tchiengue, 2833 9/2006.

### *Connarus griffonianus* Baill.
Liana; branchlets glabrescent, lenticellate; leaves 4-
jugate; leaflets oblong-obovate, to 17 × 4.7cm, apex
acuminate, base rounded to cuneate, glabrescent
beneath; petiole to 6.5cm;, petiolules 3mm; panicles to
40cm, brown pubescent; sepals valvate in bud,
triangular to ovate, 2–2.5mm, brown pubescent outside;
petals elliptic to obovate, c. 5.5 × 1.5mm, pilose
outside; follicle obliquely pyriform, c. 2 × 1.5cm,
oblique apex mucronate; stipe c. 3mm, soon
glabrescent. Forest & secondary forest; 320m.
**Distr.:** Nigeria to Angola [lower Guinea & Congolian].
**IUCN:** LC
**Bechati:** Tchiengue, 2799 9/2006.
Note: Tchiengue 2799 has immature flowers only.
Pearce, i.2010.

### *Jollydora duparquetiana* (Baill.) Pierre
Monopodial treelet, 2(–8)m; leaves imparipinnate, to
46cm; leaflets 5–11, pairs subopposite, obovate, to 30 ×
11cm, acumen 1.8cm, base acute-cuneate, glabrous;
petiole to 13.5cm; petiolules 1cm, swollen; racemes
cauliflorous, 1–several, rachis to 2cm; sepals ovate-
elliptic to oblong, to 5 × 3mm, appressed-short-hairy
outside; petals oblong, to 9 × 2mm; fruit ellipsoid, 4.5 ×
3cm, apiculate, stipitate, orange, smooth. Forest; 600m.
**Distr.:** E Nigeria to Congo (Kinshasa) [lower Guinea &
Congolian].
**IUCN:** LC
**Bechati:** Tchiengue, 2812 9/2006.
Note: Tchiengue 2812 has fruits only, Pearce, i.2010.

# CUCURBITACEAE

## L. Pearce (K)

Fl. Cameroun 6 (1967).

### *Momordica cissoides* Planch. ex Benth.
Herbaceous climber; stems thin, ridged, glabrous;
leaves palmately compound, (3–)5-foliolate c. 8 ×
5.5cm, slightly scabrid; leaflets elliptic, base acute,
assymetric in lateral leaflets, apex mucronate, margins

serrate; dioecious; male umbels and solitary female
flowers both enclosed within toothed orbicular bract (2
× 2cm) with cordate base; peduncle 4–5cm; sepals
lanceolate c. 3mm; petals elliptic 2.5 × 1cm, white-
yellow; fruits c. 3.5 × 2cm, ovate, orange, densely
bristled. Forest and farmbush; 1330–1340m.
**Distr.:** Guinea (Conakry) to Angola, E Africa [tropical
Africa].
**IUCN:** LC
**Fosimondi:** Tchiengue, 2192 4/2005; 2611 2/2006.

### *Oreosyce africana* Hook.f.
Herbaceous climber, to 3–4m; stems 1–2mm, ridged,
setose, particularly when immature; leaves (5–)8–9 ×
(4–)6–7cm, shallowly 5-lobed, base cordate, apex
mucronate, margins finely-dentate, surfaces densely
setose, hairs white; petiole 4–10cm, setose;
monoecious; flowers solitary or paired, axillary, both
male and female flowers with pubescent yellow petals,
elliptic, c. 1cm; calyx densely hairy with lanceolate
lobes to 2mm; fruits ovoid, 1–2 × 1cm, densely hispid.
Open disturbed ground, often at high elevations.
**Distr.:** Bioko & Cameroon to E & S Africa [afromontane].
**IUCN:** LC
**Fosimondi:** Tchiengue, 2168 4/2005.

### *Raphidiocystis mannii* Hook.f.
Herbaceous climber; stems thin, ridged, glabrous or
finely pubescent; leaves (3–)5-lobed, c. 11 × 7cm, base
deeply cordate, apex acuminate, margin shallowly
undulate with small mucros at apices of veins; upper
surface scabrid; petioles c. 6.5cm, pubescent; dioecious;
male flowers in branched or unbranched axillary
clusters; pedicels 0.5–2cm; female flowers solitary or
paired; sepals in both sexes strongly pinnatipartite,
individual lobes lanceolate; petals 0.7 × 0.5cm (male), 2
× 1cm (female), yellow-orange with acute apex; fruits
broadly ellipsoid, c. 4.5 × 3.5cm, densely orange-red
hairy, calyx often persistent. Forest & forest margins.
**Distr.:** Nigeria, Bioko & Cameroon [lower Guinea].
**IUCN:** NT
**Fosimondi:** Tchiengue, 2163 4/2005.

### *Zehneria scabra* (L.F.) Sond.
F.T.E.A. Cucurbitaceae: 122 (1967); Fl. Cameroun 6:
44 (1967).
**Syn. *Melothria mannii* Cogn.**
**Syn. *Melothria punctata* (Thunb.) Cogn.**
Herbaceous climber or trailer; stems 1–2mm, ridged,
scabrid around nodes; leaves 5–6(–9) × 3.5–4(–7.5)cm,
ovate, base deeply cordate, apex acute or acuminate,
margin dentate, scabrous, particularly on upper surface,
drying blackish; dioecious; male umbels c. 20-flowered;
peduncle 1–3cm, finely pubescent; pedicels 0.5cm;
female umbels 5(–10)-flowered; peduncle 1–2mm;
flowers 1–2mm; calyx pubescent, blackish; petals
white-yellow; fruit orbicular, 0.5–1cm diam., orange
when mature, finely pitted. Secondary vegetation,
farmland; 1800m.

Distr.: tropical Africa & Asia [palaeotropical].
IUCN: LC
Fosimondi: <u>Tchiengue, 2253</u> 4/2005.

# DICHAPETALACEAE

## F. Breteler (WAG)

Fl. Cameroun 37 (2001).

***Dichapetalum heudelotii*** (Planch. ex Oliv.) Baill.
var. ***hispidum*** (Oliv.) Breteler
Meded. Land. Wag. 79(16): 33 (1979).
Climber; stems 1.5mm diam., hairs 3–4mm long, dark brown, patent, moderately dense; leaves alternate, elliptic-oblong or oblanceolate-oblong, 10–17 × 3.3–5cm, acumen 1–2cm, base rounded, abruptly and shortly cordate, lateral nerves 6–8 pairs, black glands visible along midrib in transmitted light; petiole 2–5mm; stipules persistent, filiform 8mm; infructescence sessile, fasciculate; fruit fleshy, orange, ellipsoid 3 × 2cm, densely long appressed hairy. Forest; 670m.
Distr.: Cameroon & Gabon [lower Guinea].
IUCN: LC
Bechati: <u>van der Burgt, 897</u> fr., 9/2006.

***Dichapetalum tomentosum*** Engl.
Shrub 2–3m, densely green-grey puberulent; leaves elliptic or obovate-elliptic, c. 7 × 3cm, broadly acute at base and apex, c. 4–5 pairs of nerves, densely puberulent, pale grey-green below; petiole 0.6cm; inflorescence axillary, 3cm; peduncle 2cm; flowers subsessile in clusters; fruit ellipsoid, warty, 2cm. Forest; 250m.
Distr.: SE Nigeria to Gabon [lower Guinea].
IUCN: LC
Bechati: <u>van der Burgt, 850</u> fl., 9/2006.

# EUPHORBIACEAE

## B. Tchiengué (YA), M. Cheek (K) & P. Hoffmann (K)

***Alchornea floribunda*** Müll.Arg.
Shrub; leaves often subclustered, oblanceolate, to 35 × 12cm, acumen short, base cuneate then abruptly truncate, nerves 10+ pairs; petiole 1cm; male inflorescence terminal, erect, red; female inflorescence with styles 3, 2cm, red. Farmbush, secondary forest; 1700m.
Distr.: Senegal to Uganda [Guineo-Congolian].
IUCN: LC
Fosimondi: <u>Tchiengue, 1943</u> 4/2004.

***Alchornea laxiflora*** (Benth.) Pax & K.Hoffm.
Shrub to 3.5m; leaves to 12.5 × 6cm, elliptic-lanceolate to oblong-oblanceolate, acuminate or cuneate, shallowly crenate-serrate, sparingly pubescent on the midrib and main nerves at first, soon glabrescent except

in the domatia, reddish when young; petioles to 2cm; male inflorescences up to 12cm long, axillary; bracts 1.5–5 × 1–2mm, ovate; female inflorescences usually not more than 10cm long, terminal, spicate, few-flowered; bracts 2–3mm long, ovate-lanceolate; fruits 5–7 × 7–8mm, dark green, brown or black. Secondary forest and farmbush.
Distr.: Nigeria, eastwards to Ethiopia & south to South Africa [tropical Africa].
IUCN: LC
Fosimondi: <u>Tchiengue, 2264</u> 4/2005.

***Antidesma laciniatum*** Müll.Arg. var. ***laciniatum***
Tree 8–10m, puberulent; leaves obovate-oblong, 15(–30) × 8(–15)cm, subacuminate, base abruptly rounded; petiole 0.5cm; stipules ovate, 3–6-lobed, the lobes lanceolate; infructescence non-interrupted, pendulous racemose, c. 15cm; fruits ellipsoid, 7mm, red, fleshy. Forest; 1400m.
Distr.: Ivory Coast, Nigeria to Congo (Kinshasa) [Guineo-Congolian].
IUCN: LC
Fosimondi: <u>Tah 312</u> 4/2004.

***Antidesma vogelianum*** Müll.Arg.
Tree or shrub, 3–10m, pubescent; leaves elliptic, rarely slightly obovate, c. 15 × 5.5cm, acumen c. 2cm, base acute; petiole c. 5mm; stipule lanceolate, c. 6 × 2mm, entire; infructescence non-interrupted, pendulous, racemose, c. 15cm; fruits ellipsoid, 7mm, red, fleshy. Forest; 615m.
Distr.: Nigeria to Tanzania [lower Guinea & Congolian].
IUCN: LC
Bechati: <u>Tchiengue, 2827</u> 9/2006.

***Bridelia atroviridis*** Müll.Arg.
Shrub, 8–12m, inconspicuously puberulent; leaves distichous, elliptic-oblong, rarely oblanceolate-elliptic, drying black above, c. 12 × 5.5cm, acuminate, base rounded, ends of secondary nerves not joining to a marginal nerve; fruit axillary, fleshy, ellipsoid, 7mm. Semi–deciduous forest; 400m.
Distr.: Sierra Leone to Zimbabwe [tropical Africa].
IUCN: LC
Bechati: <u>van der Burgt, 873</u> fr., 9/2006.

***Bridelia speciosa*** Müll.Arg.
Shrub, or tree, to 7m, often spiny; stems dark grey-brown with raised white or brown lenticels, 1–2mm; leaves alternate, simple, elliptic-oblong, to 16 × 6cm, subacuminate, acute-obtuse, lateral nerves 12 pairs uniting in a marginal nerve, tertiary nerves scalariform; petiole 1–1.5cm; stipules translucent, 3mm; flowers sessile on leafy branches; calyx lobes 5, c. 1mm; fruit ellipsoid, fleshy, 1 × 0.5cm. Forest edges, montane grassland.
Distr.: SE Nigeria, Cameroon and CAR [Guineo-Congolian (montane)].

**IUCN:** LC
**Fosimondi:** Tchiengue, 2228 4/2005.

**Croton macrostachyus** Hochst. ex Del.
Tree, 3–8m; leaves alternate, simple, ovate, 7–14 ×
5.5–10cm, acumen 1cm, base cordate, sinus 0.5–1cm,
lateral nerves 6–8 pairs, upper surface green, dotted
with minute stellate hairs, lower surface silvery white,
with dense stellate hairs; petiole apex with a gland on
each side, globose 2mm, petiole 5–10cm, soft hairy;
inflorescence among leaves, 10cm; fruit 1.2cm.
Roadside, forest edges, cultivated; 1400m.
**Distr.:** tropical Africa.
**IUCN:** LC
**Fosimondi:** Tchiengue, 2643 2/2006.

**Discoclaoxylon hexandrum** (Müll.Arg.) Pax &
K.Hoffm.
F.T.E.A. Euphorbiaceae (1): 280 (1987); Keay R.W.J.,
Trees of Nigeria: 176 (1989); Hawthorne W., F.G.F.T.
Ghana: 132 (1990).
**Syn. *Claoxylon hexandrum*** Müll.Arg.
Pithy tree, 3–5(–10)m, minutely puberulent; leaves
elliptic, to 28 × 12cm, short acuminate, acute, margin
serrate; petiole c. 18cm; inflorescences numerous, one
per axil, pendent, spike-like, c. 12cm; flowers minute.
Forest; 1400–1800m.
**Distr.:** Liberia to Uganda [Guinea-Congolian].
**IUCN:** LC
**Fosimondi:** Tchiengue, 1937 4/2004; 2567 2/2006.

**Drypetes principum** (Müll.Arg.) Hutch.
Tree, to 6–15m; stems brown puberulent, shallowly
furrowed; leaves drying blackish, shining below,
oblanceolate, c. 15 × 6cm, acuminate, acute, petiole
8mm; stipules caducous; inflorescence fasciculate,
axillary or on leafless branches; pedicels 3mm; flowers
with sepals 5mm, pubescent; ovary pubescent, stigmas
2; fruit 15mm. Forest; 1350–1370m.
**Distr.:** Guinea (Conakry) to Cameroon [upper & lower
Guinea].
**IUCN:** LC
**Fosimondi:** Tchiengue, 2219 4/2005; 2591 2/2006.

**Macaranga occidentalis** (Müll.Arg.) Müll.Arg.
Tree, to 3–25m; trunk often spiny, with red or clear
exudate, glabrous; leaves suborbicular, 30–50cm diam.,
shallowly 3–5-lobed, lobes acuminate, base cordate,
lower surface often bluish white below and with minute
red glands; petiole c. 30cm; stipules 4 × 2cm;
inflorescence axillary, pendent, paniculate; bracts
5–10cm, deeply dentate-sinuate, densely pubescent;
female inflorescences to 18cm; fruit 1–2-lobed, 8mm.
Forest edge; 1480m.
**Distr.:** SE Nigeria & Cameroon [Western Cameroon
Uplands].
**IUCN:** NT
**Fosimondi:** Tchiengue, 2227 4/2005.

**Maesobotrya barteri** (Baill.) Hutch. var.
**sparsiflora** (Scott-Elliot) Keay
Shrub, 2–5m, glabrous; stems matt white; leaves
alternate, simple, obovate, 7–12 × 3.5–6.5, acumen
0.5cm, base acute, lateral nerves c. 5 pairs, areolae
reticulate, midrib drying orange-brown below; petioles
both long and short on same stem, 0.5–2cm, pulvinate
at base and apex; inflorescences orange, erect, axillary,
1.5–6cm, 4–10-flowered; flowers white, 2mm. Forest;
1300–1400m.
**Distr.:** Guinea to Cameroon [upper & lower Guinea].
**IUCN:** NT
**Fosimondi:** Tah 314 4/2004; Tchiengue, 1947 4/2004;
1952 4/2004.

**Mareyopsis longifolia** (Pax) Pax & K.Hoffm.
Shrub or tree, to 8–30m; no exudate; stems puberulent
at apex; leaves drying brown, oblanceolate, 52 × 15cm,
acuminate, obtuse then abruptly rounded, margin
acutely serrate-glandular, the teeth not confluent with
the brochidodromous secondary nerves, petiole
variable, 2.7cm, swollen at base and apex;
inflorescences spike-like, to 20cm, ramiflorous.
Evergreen forest; 350m.
**Distr.:** S Nigeria to Congo (Kinshasa) [lower Guinea &
Congolian].
**IUCN:** LC
**Bechati:** Tchiengue, 2765 9/2006.

**Pseudagrostistachys africana** (Müll.Arg.) Pax &
K.Hoffm. subsp. **africana**
Tree, 5–20m tall; bark whitish with green spots, slash
brown to red; leaves leathery, elliptic, 30 × 11–18cm,
subacuminate, base rounded then slightly decurrent,
with a pair of flat glands, ± serrate, nerves 21–24 pairs,
venation scalariform; petiole 2.5–5cm; stipule single,
3cm, caducous, scar completely encircling stem. Forest;
1700m.
**Distr.:** Ghana, Bioko, São Tomé, Cameroon [upper &
lower Guinea].
**IUCN:** VU
**Fosimondi:** Tchiengue, 1944 4/2004.

**Pycnocoma cornuta** Müll. Arg.
Sparsely branched shrub, 1–1.5m; terminal leaf
rosettes; leaves oblanceolate, c. 40 × 13cm, acuminate,
cuneate, decurrent, margin serrate-dentate; petiole 1cm;
inflorescences spike-like, c. 16cm, erect in upper axils,
rachis densely pubescent; bracts orbicular, villous;
female flowers c. 2cm long; ovary wings 8mm. Forest;
560m.
**Distr.:** Ghana to CAR [upper & lower Guinea].
**IUCN:** NT
**Bechati:** Tchiengue, 2809 9/2006.

**Shirakiopsis elliptica** (Hochst.) H.-J.Esser
Kew Bull. 56(4): 1018 (2001).
**Syn. *Sapium ellipticum*** (Krauss) Pax

Deciduous shrub or tree, 3–25m; white exudate, glabrous; leaves drying black, leathery, elliptic, c. 9 × 4cm, obtuse, acute, finely serrate; petiole 3mm; inflorescence terminal spike, 6cm; flowers numerous, 1mm, green; fruit globose or bilobed, fleshy, 8mm; styles 2, coiled. Forest or forest edge; 1500m.
**Distr.:** tropical & S Africa.
**IUCN:** LC
**Fosimondi:** Tah 304 4/2004.

## FLACOURTIACEAE

### Y.B. Harvey (K) & B. Tchiengué (YA)

***Dasylepis racemosa*** Oliv.
Tree, (4-)6–15m; bark smooth, green-grey with brown plaques; leaves elliptic or oblong, c. 19 × 9cm, acuminate, obtuse, undulate-dentate; petiole 1cm; inflorescence axillary, spicate, 5–6cm; flowers 1–2cm diam.; sepals 4, pink-red, 0.7cm; petals c. 8, white, stigmas 3; fruit 3-valved, globose, 1.5–2cm diam., thick-walled. Forest & *Aframomum* thicket; 1800m.
**Distr.:** SE Nigeria, Cameroon, Congo (Kinshasa) & Uganda [Guineo-Congolian (montane)].
**IUCN:** LC
**Fosimondi:** Tchiengue, 1929 4/2004.

***Oncoba dentata*** Oliv.
Keay R.W.J., Trees of Nigeria: 62 (1989); Fl. Gabon 34: 52 (1995); Adansonia (ser.3) 19: 257 (1997).
**Syn.** *Lindackeria dentata* (Oliv.) Gilg
Hawthorne W., F.G.F.T. Ghana: 131 (1990).
Tree or shrub to 15m; leaves oblong-elliptic or ovate, 15–22 × 7–12cm, shortly acuminate, dentate to subentire, lateral nerves 6–10 pairs; petioles 3–16cm; inflorescence 5–10cm; flowers in fascicles of 1–5; sepals 3, 5mm; petals 6–10, 5–10mm; fruit orange, indehiscent, with long bristles, c. 2cm diam. Forest; 1460m.
**Distr.:** Guinea (Conakry) to Sudan [Guineo-Congolian].
**IUCN:** LC
**Fosimondi:** Tchiengue, 2274 4/2005.

***Oncoba ovalis*** Oliv.
F.T.A. 1: 118 (1868); Adansonia (ser. 3) 19: 259 (1997).
**Syn.** *Camptostylus ovalis* (Oliv.) Chipp
Tree or shrub, 3–8m; leaves elliptic, c. 13 × 5cm, acuminate; petiole 3cm; inflorescence axillary, 6cm, 5–10-flowered raceme; flowers white, 1–2cm diam.; sepals 3, 5mm; petals 12, 1cm; fruit subcylindric, 4cm, orange, 6–8-winged, rostrum c. 0.5cm. Forest; 1350–1400m.
**Distr.:** Cameroon [Western Cameroon Uplands].
**IUCN:** NT
**Fosimondi:** Tchiengue, 1948 4/2004; 2605 2/2006.
Note: restricted to Mt Cameroon (8 coll.), Kupe Bakosi (19 coll.), Kongo (1 coll.), Fosimondi (2 coll.) and Nigeria (1 site) but habitat threatened only at lower edge of altitudinal range.

## GENTIANACEAE

### J. van der Waarde

***Swertia mannii*** Hook.f.
Erect herb, to 15cm; lateral branches to 8cm; leaves linear-lanceolate to narrowly elliptic to 1–3cm long; cymes more lax, 6–8-flowered; pedicels to 1.8cm; sepals lanceolate, 3mm; petals 5, lanceolate, 5.5mm, white with a purple stripe externally. Grassland; 1800m.
**Distr.:** Guinea (Conakry) to Cameroon [upper & lower Guinea].
**IUCN:** LC
**Fosimondi:** van der Waarde, 21 10/2004.
Note: this taxon really needs revising, Cheek, xi.2008.

## GERANIACEAE

### Y.B. Harvey (K)

***Geranium arabicum*** Forssk. subsp. ***arabicum***
Notes Roy. Bot. Gard. Edinburgh 42: 171 (1985).
**Syn.** *Geranium simense* Hochst. ex. A.Rich.
Straggling herb, 6–20cm; leaves orbicular in outline, 2–3cm diam., deeply palmately lobed, lobes 5, with 2–3 lateral lobes, sparingly pubescent; petiole 3–6cm; inflorescence 1–2-flowered; peduncle c. 8cm; flowers 1cm, pale pink (-white) with dark veins. Grassland-forest boundary, roadsides; 1800m.
**Distr.:** Nigeria to Kenya [afromontane].
**IUCN:** LC
**Fosimondi:** Tchiengue, 2284 4/2005.

## GESNERIACEAE

### M. Cheek (K)

Fl. Cameroun 27 (1984).

***Streptocarpus elongatus*** Engl.
Erect fleshy herb, to 1m; leaves ovate-elliptic, acuminate, rounded or subcordate at base, upper surface shortly pubescent, lower surface subglabrous; cymes lax, pedunculate; corolla tubular, white, 1cm long; fruit bright green, 5cm long, twisted, glabrous. Forest; 800m.
**Distr.:** Sierra Leone & Cameroon [upper & lower Guinea].
**IUCN:** LC
**Bechati:** van der Waarde, 10 10/2004.

***Streptocarpus nobilis*** C.B.Clarke
Epiphytic succulent herb, to 1m; leaves ovate, 12 × 7cm, base rounded to subcordate, apex acute, serrate, shortly pubescent on both surfaces; inflorescence axillary, peduncles up to 15cm, bracts 4–8mm; corolla tubular, lobes deep purple; capsule 3–6cm long, shortly pubescent with some glandular hairs. Forest; 1400m.
**Distr.:** Liberia to Cameroon, São Tomé & CAR [Guineo-Congolian].
**IUCN:** LC
**Fosimondi:** Tchiengue, 2579 2/2006.

# GUTTIFERAE

Y.B. Harvey (K), B. Tchiengué (YA) &
M. Cheek (K)

### *Allanblackia gabonensis* (Pellegr.) Bamps
Bull. Jard. Bot. Nat. Belg. 39: 347 (1969).
**Syn. *Allanblackia* sp. of F.W.T.A.**
Tree, to 10–30(–45)m; leaves obovate, c. 12 × 5cm, acuminate, obtuse to rounded, lower surface matt, lateral nerves c. 15 pairs, resin canals inconspicuous, midrib pinkish red; petiole 1.5cm, axillary cup; male inflorescence terminal, 3–15-flowered; flowers pale yellow or pink, 4.5cm diam.; petal apex rounded; staminal phalanges 5, with anthers on both upper and lower surface; central disc 5-lobed, slightly undulate; fruit ovoid, c. 15cm. Forest; 1700m.
**Distr.:** Cameroon & Gabon [lower Guinea].
**IUCN:** VU
**Fosimondi:** Tchiengue, 1941 4/2004.
Note: flowers red in the Fosimondi subpopulation.

### *Garcinia kola* Heckel
Tree, 5–12m, bole cylindrical; leaves elliptic, 13 × 6cm, acuminate, obtuse, resin canals black, parallel to nerves; petiole 1.5cm; inflorescence subumbellate, subsessile on short shoots, 10–15-flowered; flowers c. 1cm diam.; petals concave; staminal bundles 4; fruit orange, fleshy, 10cm diam. Forest edge & cultivated; 530m.
**Distr.:** Sierra Leone to Congo (Kinshasa) [Guineo-Congolian].
**IUCN:** LC
**Bechati:** Tchiengue, 2835 9/2006.
Note: Tchiengue 2835 is a unicate and the determination has not been confirmed, Harvey, ii.2010.

### *Garcinia smeathmannii* (Planch. & Triana) Oliv.
**Syn. *Garcinia polyantha* Oliv.**
Tree, 4–15m; leaves thickly leathery, drying pale brown below, narrowly oblong-elliptic, c. 20 × 8cm, obtuse, subacuminate, base obtuse, secondary nerves c. 20 pairs; petiole c. 1.5cm; inflorescence sessile, axillary, on leafy stems, umbellate-fascicled, 15–20-flowered; pedicels c. 15mm; flowers white, 1cm diam.; anthers with free filaments inserted on ligules, as long as petals; fruits globose, 2cm; stigmas 2; pedicels 4cm. Forest; 1800m.
**Distr.:** Guinea (Bissau) to Zambia [Guineo-Congolian].
**IUCN:** LC
**Fosimondi:** Tchiengue, 1917 4/2004.

### *Harungana madagascariensis* Lam. ex Poir.
Shrub or small tree, 3–6m, glabrescent; leaves ovate c. 12 × 5.5cm, acuminate, base rounded, nerves c. 10 pairs; petiole 2cm, producing bright orange exudate when broken; inflorescence a dense terminal panicle, 7–15cm; flowers white, 2mm, petals hairy; berries orange, 3mm. Farmbush; 1550m.
**Distr.:** tropical Africa & Madagascar [tropical Africa].

**IUCN:** LC
**Fosimondi:** Sight record, 7 2/2006.
Note: known only from habitat record of Tchiengue 2575.

### *Mammea africana* Sabine
Tree, to 30m; exudate yellow; glabrous; leaves opposite, simple, elliptic to long-elliptic, 10–18 × 4–6cm, sapling leaves 35 × 16cm, subacuminate, acute-obtuse, lateral nerves 15–20 pairs, each separated by 3 inter-secondaries, tertiary nerves visible to naked eye, on lower surface, reticulate, each areole with a raised swelling which in transmitted light is a translucent dot; resin canals absent; petiole 1.5cm, base lacking cup; flowers just below the leaves, single, axillary, 1–1.5cm diam; pedicel 1cm; fruit subglobose, 6cm, seeds 2–4. Forest; 1290m.
**Distr.:** Sierra Leone to Congo (Kinshasa), Angola and Uganda [Guineo-Congolian].
**IUCN:** LC
**Fosimondi:** Tchiengue, 2628 2/2006.

### *Pentadesma grandifolia* Baker f.
Bull. Jard. Bot. Brux. 35: 424 (1965).
Tree, 10–35m; buttresses to 2m; leaves elliptic-oblong or oblanceolate-oblong, 20 × 7cm, shortly acuminate, obtuse to rounded base, c. 50–70 pairs of secondary nerves, resin canals not seen, lower surface densely covered in black spots; petiole c. 2cm; inflorescence c. 6-flowered, terminal, pedicels c. 1.5cm; flowers globular, 6cm diam., white, anthers with free filaments. Forest; 1340m.
**Distr.:** Nigeria to Congo (Kinshasa) [lower Guinea & Congolian].
**IUCN:** LC
**Fosimondi:** Tchiengue, 2612 2/2006.
Note: distinguished from the better known *P. butyracea* Sabine e.g. by the presence of black dots on lower surface of the leaf blade.

### *Psorospermum aurantiacum* Engl.
Shrub or small tree, to 2(–5)m; young stems rusty-tomentose; leaves bullate, elliptic, 3.5–7 × 1.8–3.5cm, apex shortly acuminate, base acute, brown-green and sparsely tomentose above, venation impressed, densely rusty-tomentose below, obscuring the venation; petiole 4–5mm, tomentose; inflorescence terminal on lateral branches, pubescent throughout, 10–many-flowered; sepals acute, 2.5–3mm; petals cream, 4.5mm, pubescent within; fruit ovoid, wine-red. Grassland and forest edge; 1680m.
**Distr.:** Nigeria and Cameroon [Western Cameroon Uplands].
**IUCN:** VU
**Fosimondi:** Tchiengue, 2257 4/2005; 2560 2/2006.
**Local name:** Pantchou (Banwa) (Tchiengue 2257).

### *Psorospermum densipunctatum* Engl.
Shrub or small tree, to 2(–3)m; young stems sparsely pubescent; leaves densely bullate, elliptic, 3–5 ×

1.8–2.8cm, apex shortly acuminate, base acute, upper surface dark green, glossy, venation deeply impressed, lower surface pubescent only on midrib and ± on lateral nerves; petiole 3–6mm, ± pubescent; inflorescence terminal on lateral branches, cymes subumbellate, c. 10–20-flowered, puberulent throughout; sepals acute, c. 2.5mm; petals white to cream, pubescent within; fruit ovoid, wine-red. Grassland and forest edge; 1680m.
**Distr.:** Sierra Leone, Nigeria, Cameroon [upper & lower Guinea (montane)].
**IUCN:** NT
**Fosimondi:** Tchiengue, 2561 2/2006.

### *Symphonia globulifera* L.f.
Tree, to 30m, glabrous; leaves narrowly elliptic to oblong-elliptic, c. 10.5 × 2.8cm, long acuminate, acute, lateral nerves c. 30 pairs, resin canals inconspicuous; petiole 0.7cm; inflorescence terminal, c. 10-flowered; pedicels 1.5cm; flowers red, globose, c. 1cm diam.; styles ascending, aculeate, 5; fruit ovoid, 2.5cm, lenticellate; styles persistent. Forest; 1800m.
**Distr.:** tropical America, Africa & Madagascar [amphi-Atlantic].
**IUCN:** LC
**Fosimondi:** Sight record, 8 4/2004.
Note: known only from habitat notes of Tchiengue 1909–1925, iv.2004.


# HUACEAE

## M. Cheek (K)

### *Afrostyrax lepidophyllus* Mildbr.
Tree, to 30m; bole dull-white, garlic-scented, glabrous; leaves lanceolate-elliptic, c. 21 × 7cm, acumen 2cm, base rounded or obtuse, lower surface completely obscured by peltate scales, appearing pale brown or dull-white, lateral nerves c. 6 pairs; petiole 1cm; flowers axillary, 5mm; fruit ellipsoid, 2cm, fleshy. Forest; 470m.
**Distr.:** Ghana to Congo (Kinshasa) [Guineo-Congolian].
**IUCN:** LC
**Bechati:** Tchiengue, 2768 9/2006.
Note: assessed as VU by Hawthorne (1997) in ignorance of its large range in Congo (Kinshasa) where it is common and unthreatened in Equateur and Orientale provinces. It is has been reassessed as LC, Cheek, 2004.


# ICACINACEAE

## M. Cheek (K), G. Gosline & X. van der Burgt (K)

Fl. Cameroun 15 (1973).

### *Chlamydocarya thomsoniana* Baill.
Woody climber; branchlets pubescent; leaves elliptic to obovate-elliptic, 10–28 × 4–12cm, margin denticulate,

pubescent; dioecious; male flowers in catkins 3–6cm long; infructescence a spiked sphere comprising fruits 1.5cm diam., topped by an accrescent orange corolla, elongated into a 4–8cm tube, covered with bristly hairs. Forest; 450m.
**Distr.:** Sierra Leone to Congo (Kinshasa) [Guineo-Congolian].
**IUCN:** LC
**Bechati:** van der Burgt, 878 fr., 9/2006.

### *Leptaulus daphnoides* Benth.
Tree or shrub, to 15m; leaves elliptic, 6.5–16 × 2.5–6.5cm, base attenuate, base obtusely acuminate, 6–7 pairs lateral nerves ascendant, anastomosing, glabrous; inflorescence a short cyme, joined to the leaf; flowers 5-merous, white, 11 × 1.5mm, corolla tubular, linear; lobes 1mm long; anther filaments joined to the petals; style glabrous; fruit an ellipsoid drupe, 12 × 8 × 5mm, papillose. Forest; 470m.
**Distr.:** Sierra Leone to Sudan & Tanzania [Guineo-Congolian].
**IUCN:** LC
**Bechati:** van der Burgt, 859 fr., 9/2006.

### *Polycephalium lobatum* (Pierre) Pierre ex Engler
Fl. Cameroun 15: 96 (1973).
Woody climber; stems, leaves, petioles, flowers and fruit covered with erect red hairs; leaves highly variable, palmately nerved, often 3-lobed, 6–17 × 8–22cm, petiole 5–10cm; dioecious; male flowers reduced, in panicles of pedunculate heads 3–5mm diam.; female flowers in solitary axillary heads 1cm diam.; calyx tubular, 2mm; stamens 3; female calyx tubular, c. 3 × 2mm; fruits forming a hairy stellate ball; fruits 4 × 2cm, enveloped by the accrescent calyx. Forest; 620m.
**Distr.:** Sierra Leone to Congo (Kinshasa) [Guineo-Congolian].
**IUCN:** LC
**Bechati:** van der Burgt, 900 9/2006.

### *Pyrenacantha longirostrata* Villiers
Fl. Gabon 20: 85 (1973); Fl. Cameroun 15: 85 (1973).
**Syn.** *Pyrenacantha* sp. C *sensu* Keay
Climber, 5m; stem twining, minutely sparse puberulent; leaves alternate, elliptic, 10–14 × 4.5–6cm, acumen 1cm, base rounded to obtuse, lateral nerves 3 pairs, quaternary nerves raised, reticulate, with papillae in areolae; petiole 1–1.5cm, angled near base; infructescence 7cm, 0.5–1m from ground; fruits 2, bright orange, heart-shaped, 2.5 × 1.5cm; seeds dark brown, 15 × 14 × 8mm. Forest; 620m.
**Distr.:** Cameroon and Gabon [lower Guinea].
**IUCN:** EN
**Bechati:** van der Burgt, 888 9/2006.

### *Stachyanthus zenkeri* Engl.
**Syn.** *Neostachyanthus zenkeri* (Engl.) Exell & Mendonça

Woody climber, slender, twining, to 8m; young branches and petioles bristly-pilose; leaves elliptic to oblanceolate, 10–23 × 5–10cm, appressed-pubescent beneath, 6–9 lateral nerves; dioecious; flowers on long spikes in fascicles on the stem below the leaves; male spikes to 25cm; flowers 6-merous, c. 6mm; fruit a drupe, ellipsoid with 4 flat surfaces, 3 × 1.8 × 1.2cm, in neat ranks along rachis. Forest.
**Distr.:** Nigeria to Congo (Kinshasa) [lower Guinea & Congolian].
**IUCN:** LC
**Fosimondi:** Tchiengue, 2235 4/2005.

## LABIATAE

### B. Tchiengué (YA), A. Paton (K) & B.J. Pollard

### *Achyrospermum oblongifolium* Baker
Erect herbaceous, little-branched, undershrub, 30–70cm; stems little-branched, tomentose; leaves 10–18 × up to 8cm, narrowly obovate-elliptic, acuminate; inflorescence terminal, 3–5 (–10) × 2cm; bracts broadly ovate, ciliate; calyx teeth as broad as long, margin ciliate; corolla greenish-white. Forest; 320–1280m.
**Distr.:** Guinea (Conakry) to Bioko, Cameroon, São Tomé, Angola (Cabinda) [upper & lower Guinea].
**IUCN:** LC
**Bechati:** Tchiengue, 2803 9/2006; **Fosimondi:** Tchiengue, 2627 2/2006.

### *Clerodendrum globuliflorum* B.Thomas
Shrub or climber, to 3m; stems hollow; leaves large, elliptic, 15–25 × 6–12cm; inflorescences globose, dense, on older wood; calyx to 2cm long, margin ± fimbriate; corolla less than 12cm, ± glandular. Forest; 1350m.
**Distr.:** Nigeria, Bioko & Cameroon [lower Guinea].
**IUCN:** NT
**Fosimondi:** Tchiengue, 2589 2/2006.

### *Clerodendrum silvanum* Henriq. var. *buchholzii* (Gürke) Verdc.
Mem. Mus. Natl. Hist. Nat. B. Bot. 25: 1555 (1975); F.T.E.A. Verbenaceae: 110 (1992).
**Syn.** *Clerodendrum buchholzii* Gürke
**Syn.** *Clerodendrum thonneri* Gürke
Woody climber, to 10m; stems with petiolar spines; leaves elliptic or ovate, glabrous, 8–20 × 3–10cm, often confined to the canopy; inflorescence an elongate, leafless panicle, frequently cauliflorous; rachis 5–30cm long; calyx enlarged, 8–10mm; corolla white, fragrant; tube (1.5–)1.7–2.5cm; fruits red. Forest; 450m.
**Distr.:** tropical & subtropical Africa.
**IUCN:** LC
**Bechati:** van der Burgt, 879 fl., 9/2006.

### *Hyptis lanceolata* Poir.
Erect, aromatic herb; leaves lanceolate, to 11 × 3.5cm,

punctate beneath, shortly petiolate; inflorescence very dense, axillary, globose, many-flowered, c. 1.5cm across; mature calyx 4–5mm with subulate teeth; corolla white with pale mauve markings on the lip, very small. Grassland; 1460m.
**Distr.:** tropical Africa and America [amphi-Atlantic].
**IUCN:** LC
**Fosimondi:** Tchiengue, 2277 4/2005.

### *Isodon ramosissimus* (Hook.f.) Codd
Bothalia 15: 8 (1984); Fl. Rwanda 3: 311 (1985).
**Syn.** *Homalocheilos ramosissimus* (Hook.f.) J.K.Morton
Erect or straggling herb, to 4m; stems hollow, strongly quadrangular, pilose; leaves ovate, up to 7 × 4cm; inflorescence an axillary panicle of many-flowered dichotomous cymes; calyx tube declinate, ventricose, teeth subequal; corolla 5mm long, white, speckled purple in throat; upper lip very small, recurved; stamens declinate. Forest margins.
**Distr.:** Sierra Leone to Bioko, Cameroon, Sudan, Uganda, Zimbabwe [afromontane].
**IUCN:** LC
**Fosimondi:** Tchiengue, 2231 4/2005.

### *Leucas deflexa* Hook.f.
Straggling or semi-erect aromatic herb, to 2m; leaves lanceolate, cuneate at base, serrate, with an entire, acute tip, petiolate; inflorescence a densely globose axillary whorl with numerous linear-subulate bracteoles; corolla white; stamens ascending; anthers often conspicuously hairy, orange. Forest, forest margins, savanna; 1400m.
**Distr.:** Ghana, Bioko, Cameroon, Angola [Guineo-Congolian (montane)].
**IUCN:** LC
**Fosimondi:** Tchiengue, 2562 2/2006.

### *Plectranthus epilithicus* B.J.Pollard
Kew Bull. 60(1): 145 (2005).
**Syn.** *Solenostemon repens* (Gürke) J.K.Morton
Epilithic herb, to 80cm; stems annual arising from a succulent perennial tuber; leaves broadly ovate-triangular to triangular, (5–)20–55 × (5–)15– 45mm, crenate to almost lobate at maturity; inflorescence lax, up to 400 × 35mm in flower, with up to 28 verticillasters, each composed of 8–10-flowered paired cymes; mature calyx 3–4(– 4.5)mm; pedicel 4–7(9)mm; corolla mauve. Secondary forest adjacent to streams; 590–1800m.
**Distr.:** Guinea, Sierra Leone, Liberia, Ivory Coast, Ghana, Bioko, Cameroon, Gabon [Guineo-Congolian (montane)].
**IUCN:** LC
**Bechati:** Tchiengue, 2821 9/2006; **Fosimondi:** Tchiengue, 1965 4/2004.
Note: formerly known as *Solenostemon repens* (*Coleus repens* Gurke), a name never validly published. Tchiengue 1965 is tentatively placed in this taxon, Tchiengue, x.2008.

## Plectranthus insignis Hook.f.

Large, soft-wooded, monocarpic mass-flowering undershrub, from 3–5m at maturity; usually leafless at the time of flowering; inflorescence a very lax, ample racemose panicle; mature calyx 2cm long, teeth very unequal; corolla 2cm long, yellow suffused with purple; stamens declinate. Forest; 1370m.
**Distr.:** W Cameroon [Western Cameroon Uplands (montane)].
**IUCN:** NT
**Fosimondi:** Tchiengue, 2223 4/2005.
**Uses:** MEDICINES. **Local name:** Mbouh Melong (Banwa) (Tchiengue 2223).

## Plectranthus kamerunensis Gürke

Straggling, densely-woolly herb, to 1m; leaves ovate, to 11 × 9cm, acutely acuminate, cordate, coarsely crenate; inflorescences little-branched; mature calyx 8mm long, with a long white pubescence; lower teeth lanceolate, acuminate; corolla violet; stamens declinate. Forest, forest margins; 1400m.
**Distr.:** SE Nigeria, W Cameroon, E Africa [afromontane].
**IUCN:** LC
**Fosimondi:** Tchiengue, 2563 2/2006.

## Plectranthus melleri Bak.

J.Bot. 20: 243 (1882).
**Syn.** *Plectranthus luteus* Gürke
A branched woody herb, to ± 1m, producing conpsicuous fusiform, densley brown-villose bulbils in the axils of the inflorescence and branches; leaves elliptic-lanceolate, long-acuminate, cuneate, petiolate; inflorescences axillary and terminal, flowers in sessile fascicles; pedicels c. 7mm long; mature calyx glandular, c. 7mm long; corolla yellow, c. 1cm. Forest.
**Distr.:** tropical Africa and Madagascar [afromontane].
**IUCN:** LC
**Fosimondi:** Tchiengue, 2287 4/2005.

## Plectranthus occidentalis B.J.Pollard

Kew Bull. 60(1): 146 (2005).
**Syn.** *Solenostemon mannii* (Hook.f.) Baker
Herbaceous, or somewhat woody perennial herb or shrub, to c. 1m; stems climbing or erect; leaves ovate, 4–15cm, acutely acuminate, crenate, long-petiolate, often purplish-tinged; inflorescence a copiously-flowered, dense raceme, up to 25 × 3–4cm or more in fruit; calyx 4–5mm; corolla rich bluish purple. Forest, woodland; 1500m.
**Distr.:** Sierra Leone to Bioko & W Cameroon [upper & lower Guinea].
**IUCN:** LC
**Fosimondi:** Tah 311 4/2004; Tchiengue, 2269 4/2005.

## Plectranthus punctatus L'Hér. subsp. *lanatus* J.K.Morton

Erect or occasionally scrambling, glandular-pubescent herb, to 20–50cm; stems sub-succulent, swollen at the nodes, greenish-white, speckled red; leaves sessile or nearly so, up to 7.5 × 4cm; inflorescence dense in flower, lax in fruit; calyx 8–10mm long at maturity; corolla 1.2–1.5cm long, pale blue speckled purple to royal blue; stamens declinate. Damp grassland at the edges of swamps & submontane bushland.
**Distr.:** Bamboutos Mts and Bamenda highlands [W Cameroon Uplands (montane)].
**IUCN:** VU
**Fosimondi:** Tchiengue, 2170 4/2005.

## Plectranthus tenuicaulis (Hook.f.) J.K.Morton

**Syn.** *Plectranthus peulhorum* (A.Chev.) J.K.Morton
Slender, branched, annual herb, to ± 1m; stems pubescent; leaves shortly petiolate, lanceolate, acute, 0.5–6cm long, a third as broad, crenate, pubescent; inflorescence a panicle with lateral racemose branches; mature calyx 3–5mm; corolla 8–10mm, pale blue; stamens declinate. Forest, forest margins; 1800m.
**Distr.:** tropical Africa.
**IUCN:** LC
**Fosimondi:** Tchiengue, 1964 4/2004.
Note: Tchiengue 1964 is tentatively placed in this taxon by Pollard, iii.2005.

# LAURACEAE

M. Cheek (K)

Fl. Cameroun 18 (1974).

## Beilschmiedia sp. 1 of Bali Ngemba

The Plants of Bali Ngemba Forest Reserve, Cameroon: 104 (2004).
Tree, 12–25m, aromatic when wounded; stems and leaves glabrous; leaves alternate, elliptic-oblong or slightly oblanceolate, to 20 × 8cm, acumen to 2cm, base acute, midrib drying slightly yellowish brown below, finely ridged below, yellow above when live, lateral nerves 5–8pairs; petiole purple live, to 1.5cm; inflorescences dense, axillary, patent puberulent, held above branches, c. 10 × 10cm, orange-green; flowers 3–4mm wide, C3+3, A3+3, G1; fruit ellipsoid, 1-seeded, fleshy, blue-black, c. 3 × 1.5cm. Submontane forest; 1400m.
**Distr.:** Cameroon (Bali Ngemba & Fosimondi) [Cameroon Endemic].
**Fosimondi:** Tchiengue, 1953 4/2004.

# LECYTHIDACEAE

B. Tchiengué (YA)

## Napoleonaea egertonii Baker f.

Tree, to 20m; leaves oblong-elliptic, 18–36.5 × 8–13cm, apex shortly acuminate, base acute, lateral nerves 9–11 pairs; petiole to 1.2cm; panicles cauliflorous, 12 × 12cm; peduncle brown-puberulent; pedicel 0.3–1cm; calyx 2.4cm diam., 5-lobed, densely

warted and puberulous; corolla c. 4.5cm diam., white with purple veins, outer corona segments 0.45cm, puberulent, inner corona glabrous; fruit subglobose, 14 × 16.5 × 14cm, echinate, spines to 1.5cm; seeds to 5cm. Forest; 320m.
**Distr.:** S Nigeria to Gabon [lower Guinea].
**IUCN:** VU
**Bechati:** Tchiengue, 2801 9/2006.

## LEEACEAE

B. Tchiengué (YA)

Fl. Cameroun 13 (1972).

*Leea guineensis* G.Don
Erect or sub-erect, soft-wooded shrub, to 7m; leaves bipinnate; leaflets opposite, imparipinnate, oblong-elliptic to 18cm long; flowers bright yellow, orange or red; fruits brilliant red, turning black. Forest and forest gaps; 1800m.
**Distr.:** Guinea (Bissau) & (Conakry), Sierra Leone, Ivory Coast, Togo, Ghana, Nigeria, Cameroon, Equatorial Guinea, Gabon, CAR, Congo (Kinshasa) & (Brazzaville), Burundi, Sudan, Uganda, Kenya, Tanzania, Malawi, Zambia, Angola, Madagscar, Reunion, Comoros and Mauritius [tropical Africa].
**IUCN:** LC
**Fosimondi:** Tchiengue, 2258 4/2005.

## LEGUMINOSAE-CAESALPINIOIDEAE

B. Mackinder (K), X. van der Burgt (K) & B. Tchiengué (YA)

Fl. Cameroun 9 (1970).

*Duparquetia orchidacea* Baill.
Sarmentose shrub or liane, to 5m; leaves imparipinnate; leaflets, 2–4 pairs, ovate, elliptic or oblanceolate, 6–18 × 4–7cm; flowers "orchid-like", pink and white, red-veined; 3 upper petals, 11–15mm; pod compressed, lanceolate, up to 10cm, 2 prominent nerves on face of valves. Forest, stream banks; 320m.
**Distr.:** Liberia to Congo (Kinshasa) [Guineo-Congolian].
**IUCN:** LC
**Bechati:** van der Burgt, 862 fl., fr., 9/2006.

*Hymenostegia afzelii* (Oliv.) Harms
Tree, to 15m; leaves pinnate, leaf rachis winged; leaflets in 2 pairs, elliptic to ovate, upper pair larger, to 10 × 3.5cm; bracteoles 2; white or pink; sepals 4; petals 3, yellowy-green becoming pink, 6–7mm; fruit compressed, oblong-elliptic, up to 8cm. Forest; 350m.
**Distr.:** Guinea (Conakry) to Cameroon [upper & lower Guinea].
**IUCN:** LC
**Bechati:** van der Burgt, 863 9/2006.

*Leonardoxa africana* (Baill.) Aubrév. subsp. *letouzeyi* McKey
Adansonia, 22(1): 100 (2000).
Small tree, glabrous; internodes swollen, c. 4 × 0.6cm, often ant inhabited; leaves alternate, paripinnate, c. 30cm, often 4-jugate (our material juvenile at 2m, is 1 to 3-jugate); leaflets curved, elliptic, c. 20 × 7.5cm, acumen 0.5–1cm, base asymmetric, lateral nerves c. 10 pairs, uniting in loops 0.7cm from the margin; inflorescences c. 5cm; flowers dense, red, 2–3cm diam.; fruit 15 × 5cm. Secondary forest; 560m.
**Distr.:** SE Nigeria & Cameroon [lower Guinea].
**IUCN:** NT
**Bechati:** Tchiengue, 2811 9/2006.

*Scorodophloeus zenkeri* Harms
Bot. Jahrb. Syst. 30: 78 (1961).
Tree, scented of garlic when wounded; twigs puberulent, brown; leaves alternate, pinnate, 7–8cm; leaflets 4–7 on each side of rachis, alternate, suboblong, c. 2 × 0.8cm, apex asymmetrically rounded, base quadrate, petiole attached in corner, midrib diagonal, subsessile; petiole 2–10mm, pulvinate; stipules caducous; raceme 3–5cm; flowers c. 1cm diam.; calyx pink; petals white; pods c. 10 × 4cm, apex quadrate, smooth, leathery. Secondary forest; 620m.
**Distr.:** Cameroon, Gabon, Congo (Brazzaville), Congo (Kinshasa) [lower Guinea & Congolian].
**IUCN:** LC
**Bechati:** Tchiengue, 2840 9/2006.
**Uses:** FOOD ADDITIVES — Bark: bark used as spice in traditional meal (Achu) (Tchiengue 2840).

*Zenkerella citrina* Taub.
Tree or shrub, to 20m; leaves 1-foliolate; leaflet narrowly-elliptic or elliptic, 8–12 × 4.5–6cm; flowers white and pink, 5–7mm; pod broadly oblong, compressed, somewhat asymmetric, up to 7cm, leathery. Forest; 1800m.
**Distr.:** SE Nigeria to Gabon [lower Guinea].
**IUCN:** NT
**Fosimondi:** Tchiengue, 2263 4/2005.

## LEGUMINOSAE-MIMOSOIDEAE

B. Tchiengué (YA) & X. van der Burgt (K)

*Albizia adianthifolia* (Schum.) W.F.Wight
Tree, to 35m, crown flat; leaves bipinnate; leaflets numerous, sometimes hairy above, usually hairy below, occasionally hairs confined below to the mid-rib and margins, not auriculate at base, up to 2.0 × 1.1cm; inflorescence capitate; calyx and corolla inconspicuous; stamens numerous, showy, up to 2.5cm, greenish becoming red towards apex, fused into a tube, the free ends extending a further 5–7mm; pod compressed, coriaceous, finely hairy, not glossy, up to 18 × 2.6cm. Forest, farmbush; 1120m.

**Distr.:** tropical & subtropical Africa.
**IUCN:** LC
**Fosimondi:** Tchiengue, 2635 2/2006.
Note: Tchiengue 2635 is tentatively placed in this taxon (determined as *A.* cf. *adianthifolia* by Tchiengue, vi.2007).

### *Cylicodiscus gabunensis* Harms

Tree to 50m; trunk armed with spines when young; leaves bipinnate; leaflets few, glabrous, 1–2 pinnae pairs, each pinna with 5–10 alternate leaflets, up to 10.2 × 5.1cm; inflorescence a dense raceme, either solitary, in clusters or compounded into panicles, covered in stellate hairs when young; flowers inconspicuous, yellowish, up to 7mm; pod compressed, black, up to 1m × 5cm, splitting along the upper margin only. Forest; 320m.
**Distr.:** Ivory Coast to Gabon [upper & lower Guinea].
**IUCN:** LC
**Bechati:** Photographic record, 1 9/2006.
Note: known only from a photographic record (Xander van der Burgt).

# LEGUMINOSAE-PAPILIONOIDEAE

## B. Mackinder (K), R.M. Polhill (K) & B. Schrire (K)

### *Adenocarpus mannii* (Hook.f.) Hook.f.

Shrub, to 5m; leaves 3-foliolate; leaflets very variable in shape, 5–8 × 1.5–3.5cm; flowers yellow, 9–14mm; pod oblong, up to 2.5cm, viscose-glandular indumentum. Grassland & forest-grassland transition.
**Distr.:** Bioko, Cameroon, Congo (Kinshasa) & E Africa [afromontane].
**IUCN:** LC
**Fosimondi:** Sight record, 9 4/2005.
Note: known only from the habitat notes of Tchiengue 2167, 2168 & 2170.

### *Angylocalyx oligophyllus* (Baker) Baker f.

Understorey shrub, to 5m; leaves imparipinnate; leaflets (3–)5–7, alternate or subopposite, ovate, 8–25 × 3.8–8.2cm, long acuminate; inflorescence cauliflorus; flowers white with green and red or purple markings, standard petal not reflexed; pod torulose, up to 15cm, beak up to 3cm. Forest and forest gaps; 600m.
**Distr.:** Liberia to Congo (Kinshasa) [Guineo-Congolian].
**IUCN:** LC
**Bechati:** Tchiengue, 2813 9/2006.

### *Crotalaria subcapitata* De Wild. subsp. *oreadum* (Baker f.) Polhill

Polhill R., Crotalaria of Africa: 197 (1982).
**Syn.** *Crotalaria acervata* sensu Hepper
Annual or perennial erect or straggling herb, 0.5–1.3m; leaves 3-foliolate; leaflets very variable; inflorescence a raceme; peduncle shorter than rachis; flowers yellow, darkly veined, 0.5–1cm. Grassland; 1800m.
**Distr.:** tropical Africa [afromontane].
**IUCN:** LC
**Fosimondi:** van der Waarde, 14 10/2004.
Note: van der Waarde 14 is tentatively placed in this taxon as it appears to be lacking calyx indumentum, Polhill, ii.2010.

### *Crotalaria* sp. of Fosimondi

Herb, 30cm, prostrate among rocks. Submontane forest; 1800m.
**Distr.:** Cameroon (Fosimondi).
**Fosimondi:** Tchiengue, 1970 4/2004.
Note: Tchiengue 1970 has been mislaid. Generic identity was confirmed by Schrire in 2005.

### *Desmodium repandum* (Vahl) DC.

Erect herb, to 1.3m; leaves 3-foliolate; leaflets rhombic-elliptic, 4.2–9.5 × 2.8–7.5cm; flowers orange-red or red, 8–11mm; pod strongly indented along the upper margin, up to 2.5cm. Forest & forest-grassland transition.
**Distr.:** palaeotropical [montane].
**IUCN:** LC
**Fosimondi:** Tchiengue, 2174 4/2005.

### *Desmodium uncinatum* (Jacq.) DC.

Lock M., Legumes of Africa: 248 (1989).
Erect or scrambling herb, to 2m; stems with hooked hairs; leaves 3-foliolate with stipels; leaflets ovate or elliptic, 2.2–9 × 0.7–4.8cm, pubescent below; flowers pink, turning pale purple or blue, up to 1.5cm; fruits indented along both margins, up to 3cm, articles up to 3mm wide covered with hooked hairs rendering fruit 'sticky' — readily attaching to clothes. Naturalised roadside weed.
**Distr.:** native to S America, introduced elsewhere [pantropical].
**IUCN:** LC
**Fosimondi:** Tchiengue, 2230 4/2005.

### *Indigofera atriceps* Hook.f. subsp. *atriceps*

**Syn.** *Indigofera atriceps* Hook.f. subsp. *alboglandulosa* (Engl.) J.B.Gillett
Herb, to 80cm; leaves imparipinnate; leaflets 2–7 pairs plus a terminal leaflet, 8–12 × 3–5mm; inflorescence an axillary raceme; flowers deep red, 5–7mm; pod narrowly oblong, up to 12mm, covered with glandular-tipped hairs. Montane grassland; 1800m.
**Distr.:** tropical Africa [afromontane].
**IUCN:** LC
**Fosimondi:** van der Waarde, 19 10/2004.
Note: van der Waarde 19 is tentatively placed in this taxon, many diagnostic features not visible in photograph, Schrire, ii.2010.

### *Trifolium usambarense* Taub.

Straggling herb, to 1m; leaves 3-foliolate; leaflets oblanceolate, finely toothed, 6–13 × 3–7mm; calyx-nerves 10–12; flowers purple, occasionally white,

4–6mm; pod broadly oblong, c. 3 × 2mm. Marshy places & clearings in forest.
**Distr.:** Bioko, Cameroon, Congo (Brazzaville), Rwanda to E Africa, Zambia & Ethiopia [afromontane].
**IUCN:** LC
**Fosimondi:** Tchiengue, 2285 4/2005.

## LOGANIACEAE

Y.B. Harvey (K), B. Tchiengué (YA) & M. Cheek (K)

Fl. Cameroun 12 (1972).

***Anthocleista scandens*** Hook.f.
Epiphytic, climbing shrub 7–17m; branchlets square; leaves oblong-elliptic, 6–20 × 2.5–11cm; flowers white; corolla tube c. 3cm long. Forest, forest-grassland transition; 1800m.
**Distr.:** Bioko, W Cameroon & São Tomé [W Cameroon Uplands].
**IUCN:** VU
**Fosimondi:** Tchiengue, 1918 4/2004.

## LORANTHACEAE

R.M. Polhill (K) & Y.B. Harvey (K)

Fl. Cameroun 23 (1982).

***Globimetula oreophila*** (Oliv.) Tiegh.
Parasitic shrub; twigs compressed; leaves lanceolate to ovate or elliptic, 8–13 × 2.5–6cm, with 6–12 pairs of well-spaced curved-ascending nerves; umbels 1–4, in axils, 8–21-flowered; peduncle 0.5–3.5cm; corolla initially green with red upper part, darkening as bud ripens, 2.5–3.5mm; basal swelling 5-shouldered. Forest-grassland transition; 1550m.
**Distr.:** SE Nigeria & Cameroon [W Cameroon Uplands].
**IUCN:** NT
**Fosimondi:** Tchiengue, 2577 2/2006.
Note: possibly threatened by forest loss in the Bamenda Highlands.

## MALVACEAE

M. Cheek (K)

***Pavonia urens*** Cav. var. ***urens***
Subshrub; to 2m; densely persistent-pubescent on stems and leaves; leaves circular in outline, c. 15cm, more or less 5-lobed; fascicles axillary, 5–10-flowered; corolla pink, 1cm; mericarps 5; awns long exserted with retrorse spines. Forest edge; 1800m.
**Distr.:** Guinea (Conakry) to Madagascar [tropical Africa & Madagascar].
**IUCN:** LC
**Fosimondi:** Tchiengue, 2172 4/2005; van der Waarde, 17 10/2004..

## MEDUSANDRACEAE

B. Tchiengué (YA)

***Medusandra mpomiana*** Letouzey & Satabie
Adansonia (sér. 2), 14(1): 65 (1974).
Tree 10m, glabrous; leaves alternate, elliptic, c. 24 × 10cm acuminate, obtuse, 3-nerved; petiole 8cm, swollen at base and apex; inflorescence an axillary spike, 1.5cm; flowers white, 2mm; fruit globose, 2cm diam, 3-valved, 1-seeded; sepals 5, accrescent, reflexed, 0.7 × 0.5cm. Forest; 350–1150m.
**Distr.:** Cameroon [W Cameroon Uplands].
**IUCN:** NT
**Bechati:** Tchiengue, 2767 9/2006; **Fosimondi:** Tchiengue, 2618 2/2006.

## MELASTOMATACEAE

Y.B. Harvey (K), M. Cheek (K) & X. van der Burgt (K)

Fl. Cameroun 24 (1983).

***Amphiblemma mildbraedii*** Gilg ex Engl.
Erect, robust, terrestrial herb, 1–2m; stems square, to 1cm diam.; leaves ovate to 29 × 18.5cm, subacuminate, cordate, 7-nerved, subscabrid above; petiole to 10cm; stipules persistent, c. 1 × 1cm; panicle terminal, c. 18 × 15cm; peduncle 7cm; flowers 3cm, 5-petalled, purple; fruit dry, 5-angled, 0.8cm. Forest edge; 1800m.
**Distr.:** SE Nigeria, Bioko, Cameroon [W Cameroon Uplands].
**IUCN:** NT
**Fosimondi:** Tchiengue, 1907 4/2004.

***Cincinnobotrys letouzeyi*** Jacq.-Fél.
Herb, to 20cm; stem creeping, succulent, like a string of beads; leaves 2–5, clustered at stem apex, erect; blades purple below, elliptic, 5–17 × 2.5–10cm, biserrate, nerves basal, 5, both surfaces with scattered appressed white hairs; petiole 1.5–11cm; inflorescence erect; peduncle 3–14cm; flowers (1-)3–10, 2cm across; petals white with pink stripes; fruit cups to 5 × 5mm. On wet rocks in secondary forest; 460m.
**Distr.:** Cameroon (Mamfe & Fosimondi) [Cameroon Endemic].
**IUCN:** EN
**Bechati:** Tchiengue, 2807 fl., 9/2006.
Note: this is a globally very rare species being known only from the type location (Mamfe, SWP, Letouzey 14326) apart from this population on the path to Fosimondi from Bechati, Cheek, i.2010.

***Dichaetanthera africana*** (Hook.f.) Jacq.-Fél.
**Syn.** *Sakersia africana* Hook.f.
Shrub to small tree, c. 6–12m tall; leaves elliptic, strigose, apex acuminate; petiole c. 2cm long; inflorescence terminal panicle; flowers 4-merous; hypanthium and calyx lobes glabrous; buds c. 0.5cm

long, petals pink; fruit a capsule. Secondary bush, lowland rainforest; 1650m.
**Distr.:** Sierra Leone to Angola [upper & lower Guinea].
**IUCN:** LC
**Fosimondi:** Tchiengue, 2641 2/2006.

### *Dinophora spenneroides* Benth.
Shrubby herb, 1.5m, glabrous; leaves ovate-oblong, 5–14 × 3–6cm, membranous, acuminate, cordate, 5–7-nerved at base; inflorescence lax, terminal panicles; calyx dentate, white; petals 1cm, pink; berries white. Forest edge; 320m.
**Distr.:** Guinea (Conakry) to Congo (Kinshasa) & Angola [Guineo-Congolian].
**IUCN:** LC
**Bechati:** Tchiengue, 2802 9/2006.

### *Dissotis bamendae* Brenan & Keay
**Syn.** *Dissotis princeps* (Kunth) Triana var. *princeps auctt. non*
Fl. Cameroun 24: 29 (1983).
Erect herb, 0.5m, setose; leaves 3 per node, ovate, c. 9 × 4.5cm, acute, base rounded, felted below; petiole 1.5cm; flowers c. 20; peduncle 10cm; hypanthium 10 × 7mm, densely white-simple-hairy, emergences nil; sepals caducous; stamens dimorphic; appendages 2, globose. Grassland; 1800m.
**Distr.:** Cameroon [W Cameroon Uplands].
**IUCN:** VU
**Fosimondi:** Tchiengue, 1932 4/2004.
Note: sunk into *D. princeps* by Jacques-Felix (South Africa to Ethiopia) but obviously distinct in e.g. 3 leaves per node.

### *Medinilla mirabilis* (Gilg) Jacq.-Fél.
Fl. Cameroun 24: 115 (1983).
**Syn.** *Myrianthemum mirabile* Gilg
Stem twining liane; leaves in whorls of 3, oblanceolate or elliptic, to 22 × 9cm, subacuminate, cuneate-obtuse, 3-nerved; petiole to 3cm; inflorescence cauliflorous, near ground, fascicles of c. 40 flowers; peduncles to 2cm, much branched; pedicels 1.5cm; flowers 1cm, blue; berries 1cm. Forest; 615m.
**Distr.:** SE Nigeria, Bioko, Cameroon & Gabon [lower Guinea].
**IUCN:** NT
**Bechati:** van der Burgt, 892 fr., 9/2006.

## MELIACEAE

M. Cheek (K), B. Tchiengué (YA), B. Oben & Y.B. Harvey (K)

### *Carapa grandiflora* Sprague
Tree, 6–20m, glabrous; leaves to 1.2m, paripinnate, 4–7-jugate; petiole c. 15cm; leaflets oblong to oblong-obovate, c. 18 × 7cm, rounded, acute; petiolules c. 1cm; inflorescence a terminal panicle, c. 30cm; flowers white, c. 8mm; sepals and petals greenish; staminal tube white;

disc orange; stigma white; fruit 5-valved, subglobose, c. 10cm, warty; seeds c. 3cm. Forest; 1800m.
**Distr.:** Nigeria to Uganda [lower Guinea & Congolian (montane)].
**IUCN:** NT
**Fosimondi:** Tchiengue, 1925 4/2004.
Note: *Carapa* is currently being revised, after which several additional taxa, further to the two accepted in FWTA, might be revealed, Onana, vi.2009.

### *Carapa procera* DC.
Tree, c. 15m; leaves variable, to 2m, 6–21-jugate; leaflets to 50 × 15cm; panicles to 80cm; flowers 8mm. Forest.
**Distr.:** S America & Senegal to Congo (Kinshasa) [amphi-Atlantic].
**IUCN:** LC
**Fosimondi:** Sight record, 14 4/2004; 2/2006.
Note: known only from habitat notes taken iv.2004 and ii.2006, no collections made, Harvey, ii.2010.

### *Guarea glomerulata* Harms
Shrub or small tree, 1.8–4(–6)m; leaves c. 5-jugate, rachis 30cm; petioles 15cm, terete, puberulent; leaflets elliptic or oblanceolate-elliptic, c. 24 × 8cm, long acuminate, acute, lateral nerves c. 10 pairs; inflorescence a pendulous spike, c. 0.5m; flowers pink, scattered on distal part; fruit subglobose, rostrate, 1.5cm, densely brown-hairy. Forest; 1250m.
**Distr.:** Nigeria to Congo (Kinshasa) [lower Guinea and Congolian].
**IUCN:** LC
**Fosimondi:** Tchiengue, 2624 2/2006.

### *Heckeldora ledermannii* (Harms) J.J. de Wilde
Blumea 52(1): 184 (2007).
Shrub or small tree, to 5m; leaves 5–13-foliolate; petiole 4.5–15cm long; rachis (4-)16–29cm long; leaflets c. 10–20 × 2–7cm, acuminate, inflorescences raceme like, to c. 70cm long; pedicel c. 1.5mm long; calyx tiny, <1 × 2mm, puberulous; flower 6–7 × 2–2.5mm, ovary glabrous; fruit (immature) obovoid, 2–3 × 1–2cm, dull grey-green, puberulous. Transition zone of evergreen tropical rain forest of low and medium altitudes towards submontane forest; 1370m.
**Distr.:** Cameroon (SW province) [Cameroon endemic].
**IUCN:** EN
**Fosimondi:** Tchiengue, 2220 4/2005.

### *Heckeldora staudtii* (Harms) Staner
**Syn.** *Heckeldora latifolia* Pierre
Shrub or small tree, 2.5(–3))m, sparsely puberulous; leaves c. 6-jugate, rachis to c. 30cm; petiole 2.5–16cm, terete; leaflets oblong-elliptic, to 20 × 6cm, acuminate, acute, lateral nerves c. 12 pairs; inflorescences pendulous spikes to 40cm; flowers white, 6mm; calyx 1–2(–3) × 2–3mm; ovary glabrous or pubescent; fruit ovoid to cylindrical, to c. 2.5–6 × 1.5–3cm, pinkish-brown or orange-yellow, c. 2–4(–5)-seeded, indehiscent, fleshy. Evergreen, semi-deciduous lowland

forest and secondary forest; 580m.
**Distr.:** S Nigeria, Cameroon, Equatorial Guinea, Gabon, Congo (Brazzaville), Congo (Kinshasa) [lower Guinea & Congolian].
**IUCN:** LC
**Bechati:** <u>van der Burgt, 883</u> fr., 9/2006.
Note: van der Burgt 883 has subglobose fruits.

*Turraea vogelii* Hook.f. ex Benth.
Woody climber, to 12m, rarely a shrub to 2m, glabrous; leaves simple, elliptic, c. 11 × 5cm, subacuminate, obtuse; petiole 0.5cm; inflorescence subumbellate, axillary; peduncle 3.5cm; pedicels 1cm; flowers white, 3–10; staminal tube 1.5cm; fruit globose, 2.5cm. Forest; 320m.
**Distr.:** Ghana to Uganda [Guineo-Congolian].
**IUCN:** LC
**Bechati:** <u>Tchiengue, 2777</u> 9/2006.

## MELIANTHACEAE

M. Cheek (K)

*Bersama abyssinica* Fresen.
**Syn.** *Bersama maxima* Baker
**Syn.** *Bersama acutidens* Welw. ex Hiern
Tree or shrub, 2–8m, glabrous; leaves alternate, c. 30cm, variable, imparipinnate; leaflets 5–6 pairs, densely pubescent or glabrous, glossy, oblong-elliptic, c. 15 × 5.5cm, apex acute, base obliquely obtuse, lateral nerves c. 10 pairs, sometimes serrate in upper half; petiolule 0.5cm, rachis more or less winged; petiole c. 12cm; stipules c. 1cm, intrapetiolar; inflorescence a terminal raceme to c. 40cm; rachis c. 7cm; pedicels 1cm; flowers white, 1cm; fruit magenta red, dehiscent, ovoid, 2cm; seeds 1cm, arillate. Forest; 400–1400m.
**Distr.:** tropical Africa [afromontane].
**IUCN:** LC
**Bechati:** <u>van der Burgt, 876</u> fr., 9/2006; **Fosimondi:** <u>Tchiengue, 2578</u> 2/2006.
Note: we follow Verdcourt in treating all the African members of the genus as one variable species.

## MENISPERMACEAE

Y.B. Harvey (K), B. Tchiengué (YA) & X. van der Burgt (K)

*Penianthus longifolius* Miers
Shrub, to 3(–4.5)m; leaves obovate to elliptic, 15–34(–40) × 6–15(–18)cm, rounded to cuneate; midrib slightly, gradually prominent above; petiole 5–18cm; flowers remaining almost closed; stamens not protruding at maturity, to 2.8mm; fruit (1.5–)1.8–3.1(–3.4) × 1.1–1.8cm. Forest; 1340–1400m.
**Distr.:** Nigeria, Cameroon, Bioko, Rio Muni, Gabon, CAR, Congo (Brazzaville) & (Kinshasa), Angola (Cabinda) [Guineo-Congolian].
**IUCN:** LC
**Fosimondi:** <u>Tchiengue, 1951</u> 4/2004; <u>2601</u> 2/2006.

*Perichasma laetificata* Miers var. *laetificata*
Adansonia (ser.2) 17: 223 (1977).
**Syn.** *Stephania laetificata* (Miers) Benth.
Climber, to 15m; stem striate/ribbed, hispid; petiole ribbed, hispid, 9–15cm, inserted 4–6cm from base of lamina, base pulvinate and twisted, drying black; lamina peltate, to 27 × 22cm, usually palmately 9-nerved, chartaceous, margin ciliate; × inflorescence a very large panicle, to 50cm, axillary, with shortly-pedicellate lateral racemose branches, dimishing in size from base to apex. Forest; 350m.
**Distr.:** Nigeria, Cameroon, Bioko, Congo (Kinhasa), CAR, Gabon, Angola [lower Guinea & Congolian].
**IUCN:** LC
**Bechati:** <u>van der Burgt, 864</u> fr., 9/2006.

## MONIMIACEAE

B. Tchiengué (YA)

Fl. Cameroun 18 (1974).

*Xymalos monospora* (Harv.) Baill. ex Warb.
Shrub or small tree, 3–8(–25)m; leaves opposite, leathery, elliptic, c. 10 × 4cm, acute, serrate; inflorescences c. 4cm, below leaves; fruit elliptic, 1cm with apical knob. Forest & forest-grassland transition; 1800m.
**Distr.:** SE Nigeria to E & S Africa [afromontane].
**IUCN:** LC
**Fosimondi:** <u>Tchiengue, 1904</u> 4/2004.

## MORACEAE

Y.B. Harvey (K) & M. Cheek (K)

Fl. Cameroun 28 (1985).

*Dorstenia barteri* Bureau var. *barteri*
Herb, to 90cm; stem ascending, sometimes branched; leaves in spirals or sub-distichous; lamina variable, lanceolate to elliptic, broadest above, narrowed below, (3–)6–22 × (1.5–)3–8.5cm, usually entire, but sometimes with 1–2 (–5) blunt teeth in the upper third; receptacle with c. 5 or 6 primary appendages, to 3cm, secondary to 0.7cm; fringe 1–5(–10)mm. Forest; 1300–1540m.
**Distr.:** SE Nigeria, Bioko, W Cameroon & Congo (Kinshasa) [lower Guinea & Congolian].
**IUCN:** LC
**Fosimondi:** <u>Tah 319</u> 4/2004; <u>Tchiengue, 1946</u> 4/2004; <u>2570</u> 2/2006; <u>2588</u> 2/2006; <u>2614</u> 2/2006.

*Dorstenia barteri* Bureau var. **nov. of Bechati**
Erect herb, 0.3–0.5m; exudate white, stem with long white appressed hairs; leaves alternate; blade oblanceolate, c. 18 × 6–7cm, acumen 1cm, base cuneate, abuptly rounded, lateral nerves c. 8 pairs; petiole 1–1.5cm; inflorescences green, axillary, 2 per axil, square in plane view, 1.5–1.8cm wide, including

the four acuminate corners; flange 3–4mm wide; fertile part orbicular, 6–7mm wide; receptacle nearly flat; peduncle 2–2.5cm. Forest; 320–560m.
**Distr.:** Cameroon (Bechati).
**Bechati:** <u>Tchiengue, 2788</u> 9/2006; <u>2810</u> 9/2006.

### *Ficus ardisioides* Warb. subsp. *camptoneura* (Mildbr.) C.C.Berg
Fl. Cameroun 28: 238 (1985); Kew Bull. 43: 77 (1988); Keay R.W.J., Trees of Nigeria: 297 (1989); Hawthorne W., F.G.F.T. Ghana: 122 (1990); Kirkia 13(2): 271 (1990).
**Syn.** *Ficus camptoneura* Mildbr.
Epiphytic shrub, to 6m, occasionally a tree, to 12m; epidermis flaking, glabrous; leaves obovate, elliptic, or elliptic-oblong, 9–22 × 4–9cm, acuminate, obtuse, lateral nerves c. 4 pairs, basalmost arising 0.5cm or more above leaf base, quaternary nerves conspicuous; stipules subpersistent, 9mm; figs axillary, sessile, globose 0.7–1cm diam., often verrucose, subrostrate, basal bracts 2, 4mm. Forest; 1120m.
**Distr.:** Ivory Coast, Nigeria to E Congo (Kinshasa) & N Zambia [Guineo-Congolian].
**IUCN:** LC
**Fosimondi:** <u>Rønsted 237</u> 4/2004.

### *Ficus* sp. of Fosimondi
Shrub (or tree sapling), to 2m, glabrous; exudate white; leaves alternate, simple, stipules leaving annular scars; blade elliptic, 6–12 × 3–6cm, including acumen 1–1.5cm, base obtuse-acute, drying black above with dark red nerves, below orange-brown, lateral nerve pairs 2, subopposite inserted c. 0.5 and 2cm from the leaf-base, distal veins running most of the length of the blade, tertiary nerves conspicuous; petiole 0.7–1.5cm. Forest.
**Distr.:** Cameroon (Fosimondi).
**Fosimondi:** <u>Tchiengue, 2207</u> 4/2005.
Note: Tchiengue 2207 (shrub to 2m) has not been matched with other specimens at K, although very distinctive — just 2 pairs of secondary nerves, basal, very strong. This may represent a new species to sceince but requires fertile material, and checking by World Authority C.C. Berg, Cheek, i.2010.

## MYRSINACEAE

Y.B. Harvey (K), B. Tchiengué (YA) & M. Cheek (K)

### *Ardisia kivuensis* Taton
Bull. Jard. Bot. Nat. Belg. 49: 102 (1979).
Shrub, 0.3–2m; leaves ovate-elliptic or oblong, 13–18cm; petioles pink; flowers purple or reddish, in small clusters along branches; fruits bright red. Forest; 1700m.
**Distr.:** Cameroon, Congo (Kinshasa) to Uganda [lower Guinea & Congolian].
**IUCN:** LC
**Fosimondi:** <u>Tah 293</u> 4/2004.

### *Ardisia staudtii* Gilg
Bull. Jard. Bot. Nat. Belg. 49: 112 (1979).
**Syn.** *Afrardisia cymosa* (Bak.) Mez
**Syn.** *Afrardisia staudtii* (Gilg) Mez
Shrub, (0.5–)1.5–4(–5)m tall, glabrous; leaves elliptic to ovate, 90–180 × 30–70mm, glandular dots present on lower surface, very shallowly crenate; petioles 7–15mm; flowers in axillary fascicles; peduncles 2–5mm; 6–12 flowers per fascicle; pedicel 6–10mm; calyx c. 2.5mm wide, fimbriate margin; flowers white or pink, to 4mm with glandular spots/streaks; fruits globose, 3–6.5mm, red with red gland-dots. Lowland & submontane forest; 1370–1800m.
**Distr.:** Nigeria to Congo (Kinshasa) & CAR [Guineo-Congolian (montane)].
**IUCN:** LC
**Fosimondi:** <u>Tchiengue, 1963</u> 4/2004; <u>2221</u> 4/2005.
Note: Tchiengue 2221 is tentatively placed in this taxon by Tchiengue, vii.2007.

### *Embelia schimperi* Vatke
Woody climber, straggling over shrubs, to about 2m; stems woody, brown, immature growth paler; bark fissured, lenticels pale; leaves elliptic (to obovate), 4.5–10.5 × 2–4.7mm, coriaceous, obtuse, with tiny gland-dots, otherwise glabrous, entire, slightly revolute, lower surface paler than upper (almost silvery); petioles 1–2cm; raceme 2–5.5cm, on leafless part of previous year's branchlet; calyx 1.5–3.5mm in diam.; pedicels c. 3mm, hairs chestnut; flowers whitish, 4–5-lobed, 5–6.5mm diam.; fruits red, globose, 5–8mm, 1-seeded. Forest-grassland transition; 1300m.
**Distr.:** Cameroon to Ethiopia [afromontane].
**IUCN:** LC
**Fosimondi:** <u>Tah 310</u> 4/2004.

### *Maesa lanceolata* Forssk.
Tree or shrub, 6–8m tall; stem glabrous, dark brown with paler lenticels; leaves elliptic, 9–16 × 3.5–6cm, serrulate, glabrous; petioles 2–2.5cm, glabrous; inflorescence many branched, 5–7mm, profusely covered with minute hairs (<0.5mm); flowers pale green, to 1.5mm, subsessile; fruits globose, 4–5mm, pedicel to 3.5mm. Forest-grassland transition, montane forest; 1800m.
**Distr.:** Guinea (Conakry) to Madagascar [afromontane].
**IUCN:** LC
**Fosimondi:** <u>Sight record 11</u> 4/2004; 4/2005.
Note: known only from sight records in habitat notes (Tchiengue 1903–1908, iv.2004 & 2253–2254, iv.2005), Harvey, ii.2010.

## MYRTACEAE

Y.B. Harvey (K)

### *Eucalyptus* sp.
Tree, 10–20m; bole pink-brown; leaves narrowly lanceolate, slightly falcate, c. 12 × 2.5cm, apex long-attenuate, base unequally obtuse, midrib pale below,

lateral nerves numerous, inconspicuous; infloresences 5–7cm; partial-peduncles flattened; flowers white, c. 5mm diam.; calyx c. 6 × 3mm, including slightly narrower stipe. Cultivated.
**Distr.:** native to Australia [widely cultivated].
**Fosimondi:** Photographic record, 3 4/2005.
Note: many species of *Eucalyptus* have been introduced into Africa and elsewhere from E Malesia (few) and Australia. The description above is meant to be a general one to cover the various species that may occur in our area. Known only from a photographic record, Harvey, ii.2010.

### *Syzygium staudtii* (Engl.) Mildbr.
Tree, 8–20m; bole white, usually with 3-numerous laterally-flattened root buttresses, arising up to 60cm above ground; stems near apex red when young, 4-ridged, glabrous; leaves (fruiting stems) elliptic 6–7 × 2–3.5cm, acute, secondary nerves numerous; petiole c. 1.2cm; juvenile leaves to 11 × 5cm, sometimes briefly acuminate; inflorescence terminal, 10cm, 10–30-flowered; flowers white, 0.6cm; fruit subumbellate, obovoid, 1cm. Forest; 1400m.
**Distr.:** Liberia to Cameroon [upper & lower Guinea (montane)].
**IUCN:** NT
**Fosimondi:** Tah 297 4/2004; 298 4/2004.
Note: unjustifiably reduced to a subspecies of *S. guineense* by White. Likely to rate as VU when taxon better delimited.

## OLACACEAE

G. Gosline

Fl. Cameroun 15 (1973).

### *Olax gambecola* Baill.
Shrub, to 3m; branches 4-winged, covered with whitish dots; leaves ovate-elliptic to lanceolate, 7–15 × 3–6cm, petiole 0–2mm; inflorescence an axillary raceme 1–2.5cm long; flowers 2.5–3mm long, white, stamens 3, staminodes 5–6; fruits red, calyx a disc at base, not accrescent enveloping fruit. Forest, especially by streams; 520m.
**Distr.:** Guinea (Conakry) to Uganda [Guineo-Congolian].
**IUCN:** LC
**Bechati:** van der Burgt, 890 9/2006.

### *Olax latifolia* Engl.
Shrub, to 3m; branches 4-sided; leaves variable, elliptic to oblong, 7–22 × 3–10cm, 5–12 lateral nerves, petiole 0–4mm long; flowers pink, c. 4mm long, stamens 6, staminodes 3, in axillary racemes; fruits globose, 2cm diam., completely enveloped by accrescent calyx. Forest; 650m.
**Distr.:** Cameroon to Congo (Kinshasa) [lower Guinea & Congolian].

**IUCN:** LC
**Bechati:** van der Burgt, 898 9/2006.

## OLEACEAE

M. Cheek (K)

### *Jasminum preussii* Engl. & Knobl.
Woody climber, to 5m; densely brown-pubescent on young stems, leaves, pedicels and calyx, hairs to 2mm; leaves papery, sub-opposite borne on lateral branches, ovate, 5.7–6.1 × 2.5–3.1cm, apex acuminate, base rounded or sub-cordate, petiole 0.2–1.5cm; flowers solitary, terminal on lateral branches; pedicel 1.5cm in fruit, thickened to 1.5mm at apex; calyx lobes linear, 0.6cm, corolla white; fruits globose, 0.9cm diam., white, glabrous. Forest & farmbush; 1450m.
**Distr.:** Ghana to Cameroon [upper & lower Guinea].
**IUCN:** NT
**Fosimondi:** Tchiengue, 2205 4/2005.
Note: Breteler has redelimited this species to include glabrous specimens previously determined as *Jasminum pauciflorum*, extending the range and frequency of the taxon from Guinea to Cameroon (his dets at K), Cheek, xi.2008.

## OXALIDACEAE

Y.B. Harvey (K)

### *Biophytum talbotii* (Baker f.) Hutch. & Dalziel
Unbranched herb, gregarious, 10–15cm, pubescent; leaves whorled, paripinnate, 7–10cm, leaflets 10–20 pairs, median leaflets obliquely rectangular, 8–10 × 3mm, acute, glabrous; peduncle to 2mm; bracts to 7mm; flowers pink or purplish-yellow, c. 1cm; stamens yellow. Shady river banks; 250m.
**Distr.:** Liberia, Nigeria, S Cameroon, Congo (Kinshasa) [upper & lower Guinea].
**IUCN:** NT
**Bechati:** Tchiengue, 2760 9/2006.

## PANDACEAE

M. Cheek (K)

Fl. Cameroun 19 (1975).

### *Microdesmis* cf. *puberula* Hook.f.
Shrub, 2(–8)m, densely puberulent, apical bud clawed; leaves alternate, elliptic, c. 10 × 4cm; inflorescences axillary fascicles of c. 10 florets; flowers orange or pink, flat, c. 3mm diam.; fruit globose 4mm, orange, tuberulate endocarp. Forest; 350m.
**Distr.:** Nigeria to Uganda [lower Guinea & Congolian].
**Bechati:** Tchiengue, 2762 9/2006.
Note: western Cameroonian material of this genus needs critical investigation.

# PENTADIPLANDRACEAE

M. Cheek (K)

Fl. Cameroun 15 (1973).

***Pentadiplandra brazzeana*** Baill.
Tree, or climber, 20m; leaves alternate, elliptic 12.5 ×
4cm, acuminate, 1.5cm; petiole 1cm; inflorescence
axillary, 0.7cm, few-flowered; flowers 3cm, white
mottled purple; calyx, corolla and androecium free;
fruit ovoid c.8 × 7cm, hard, mottled-grey, pointed,
flesh-orange; seeds 2cm, white. Forest; 1400m.
**Distr.:** Cameroon to Congo (Kinshasa) [lower Guinea
& Congolian].
**IUCN:** LC
**Fosimondi:** Tchiengue, 1958 4/2004.

# PHYTOLACCACEAE

Y.B. Harvey (K)

***Phytolacca dodecandra*** L'Hér.
Dioecious shrub, sometimes scandent, to 4m, glabrous,
slightly fleshy; leaves alternate, ovate-elliptic, c. 8 ×
3cm, acute, base rounded; petiole c. 2cm; inflorescence
racemose, c. 20cm; pedicels 0.6cm; sepals 4, petals absent;
stamens c. 15; berries red, c. 0.8cm. Forest; 1800m.
**Distr.:** tropical & subtropical Africa.
**IUCN:** LC
**Fosimondi:** Tchiengue, 1935 4/2004.

# PIPERACEAE

M. Cheek (K) & B. Tchiengué (YA)

***Peperomia fernandopoiana*** C.DC.
Epiphytic herb, c. 4m from ground; stems erect,
branched, 30cm, drying black, glabrous; leaves alternate,
ovate-lanceolate, c. 7 × 3.5cm, acumen long, acute;
inflorescences terminal and axillary, 2–3 per peduncle, to
6cm. Forest, secondary forest beside water; 1800m.
**Distr.:** Sierra Leone to Kenya [tropical Africa].
**IUCN:** LC
**Fosimondi:** Tchiengue, 1923 4/2004.

***Piper capense*** L.f.
Pithy shrub, c. 1(–5)m; peppery aroma emitted when
crushed; leaves opposite, broadly ovate, c. 15 × 10cm,
cordate, glabrous except on nerves; inflorescences of
leaf opposed, single, erect, white spikes c. 3 × 0.5cm.
Forest; 1800m.
**Distr.:** Guinea (Conakry) to South Africa [tropical &
subtropical Africa].
**IUCN:** LC
**Fosimondi:** Tchiengue, 1908 4/2004.

***Piper guineense*** Schum. & Thonn.
Hemiepiphyte-climber, reaching 20m above ground;
peppery when crushed; stem twining and rooting

adventitiously; leaves ovate-elliptic, to 19 × 10cm,
obliquely-obtuse at base; inflorescence single, leaf-
opposed, 3cm. Forest; 590m.
**Distr.:** Guinea (Bissau) to Uganda [Guineo-Congolian].
**IUCN:** LC
**Bechati:** Tchiengue, 2823 9/2006.
Note: Tchiengue 2823 is a unicate and the identification
is unconfirmed, Harvey, ii.2010.

# PITTOSPORACEAE

Y.B. Harvey (K)

***Pittosporum viridiflorum "mannii"*** Sims
Meded. Land. Wag. 82(3): 260 (1982); Kew Bull. 42:
328 (1987).
**Syn.** *Pittosporum viridiflorum* Sims subsp. ***dalzielii***
(Hutch.) Cuf.
**Syn.** *Pittosporum mannii* Hook.f.
Shrub or small tree, to 10m, immature branches and
petioles pubescent to glabrescent; leaves usually
crowded towards the end of the branches, c. 7–17 ×
1.5–4cm, lanceolate to spathulate, acuminate, cuneate;
inflorescences paniculate, with pubescent branches;
pedicels to 5mm; sepals to 1.2 × 0.8mm, glabrous or
ciliolate, free or basally connate; petals to 5mm; capsule
valves to 8mm diam.; seeds mostly 4 on each valve.
Forest, bushland and occasionally farmland; 1800m.
**Distr.:** Cameroon, Nigeria and Bioko [Guineo-Congolian].
**IUCN:** NT
**Fosimondi:** Tchiengue, 1933 4/2004.

# POLYGALACEAE

M. Cheek (K)

***Heterosamara cabrae*** (Chodat) Paiva
Fontqueria 50: 128 (1998).
**Syn.** *Polygala cabrae* Chodat
Decumbent herb, or subshrub, to 60cm; stems
pubescent; leaves elliptic, 3.3–11 × 1.4–5cm, apex
acuminate-mucronate, base attenuate, margin obscurely
undulate, upper surface and veins of lower surface
setulose, especially in young leaves; petiole to 1.4cm;
racemes axillary or terminal, lax; rachis to 6cm,
pubescent; lateral sepals oblong, 0.9cm, purple-white,
remaining sepals ovate, 3mm; petals 3, purple-white;
fruit to 4mm, emarginate, narrowly-winged. Forest &
forest edge; 1300–1400m.
**Distr.:** Cameroon to Congo (Kinshasa) [lower Guinea].
**IUCN:** NT
**Fosimondi:** Tah 299 4/2004; Tchiengue, 1955 4/2004.

# RANUNCULACEAE

M. Cheek (K)

***Thalictrum rhynchocarpum*** Quart.-Dill. &
A.Rich. subsp. ***rhynchocarpum***

Herb, c. 60cm, glabrous; leaves tripinnate, c. 20cm; leaflets orbicular-elliptic, c. 1.2 × 1.2cm, entire or 3–5-lobed, petiolules capillary; inflorescence c. 30cm; flowers numerous, c. 0.4cm, green. Forest edge; 1800m.
**Distr.:** Bioko to Tanzania & South Africa [afromontane].
**IUCN:** LC
**Fosimondi:** <u>Tchiengue, 2255</u> 4/2005.

## RHIZOPHORACEAE

Y.B. Harvey (K)

*Cassipourea malosana* Alston
Kew Bull. 1925: 258 (1925).
Tree, 3m; leaves opposite, elliptic, 9 × 3.5cm, including 1.5cm acumen, base cuneate, serrate in upper $2/3$, lateral nerves c. 6 pairs; petiole c.1cm; interpetiolar stipule triangular, sericeous, 6 × 3mm, rounded; flowers 1–3, axillary-fasciculate, c. 4 fertile nodes per stem; pedicel 7mm; flower 8mm. Forest; 1700m.
**Distr.:** Cameroon, Ethiopia, Somalia, Uganda, Kenya, Tanzania, Malawi, Mozambique, Zambia, Zimbabwe. [afromontane].
**IUCN:** LC
**Fosimondi:** <u>Tah 287</u> 4/2004.
Note: further research may well show that the Cameroon montane population merits specific distinction from the E African populations. Likely to be confused with Rubiaceae but for the toothed leaves and superior ovary.

## ROSACEAE

B. Tchiengué (YA)

*Prunus africana* (Hook.f.) Kalkman
F.T.E.A. Rosaceae: 46 (1960); Blumea 13: 33 (1965); Fl. Cameroun 20: 209 (1978); Fl. Ethiopia 3: 32 (1989); Keay R.W.J., Trees of Nigeria: 181 (1989).
**Syn.** *Pygeum africanum* Hook.f.
Tree, to c. 20m; leaves alternate, lanceolate, 3–6 × 6–15cm, serrate; petiole 2cm long, bearing 2 glands near apex, or at base of lamina; inflorescence a dense panicle; flowers white, 5mm diam.; fruit a drupe, succulent, red, c. 1cm diam. Forest-grassland transition; 2150m.
**Distr.:** tropical and subtropical Africa [afromontane].
**IUCN:** LC
**Fosimondi:** <u>Sight record, 13</u> 11/1974.
Note: known from Letouzey habitat notes (No. 13382 & 13386, xi.1974), no collections made, Harvey, ii.2010.

*Rubus apetalus* Poir.
**Syn.** *Rubus exsuccus* Steud. ex A.Rich.
**Syn.** *Rubus rigidus* Sm. var. *camerunensis* Letouzey
Fl. Cameroun 20: 226 (1978).
Spiny shrub; stems densely pubescent; leaves 3-foliolate, pubescent below; petals 2–3, small, caducous; fruits ripening black. Forest-grassland transition; 1800m.
**Distr.:** tropical and S Africa [afromontane].
**IUCN:** LC
**Fosimondi:** <u>Tchiengue, 1926</u> 4/2004.

## RUBIACEAE

M. Cheek (K), S. Dawson (K), B. Tchiengué (YA) & O. Lachenaud (BR)

*Argocoffeopsis fosimondi* Tchiengué & Cheek
The Plants of Lebialem Highlands, Cameroon: 54 (2010).
Shrub 1.5–3m, glabrous; leaves elliptic-lanceolate, 14–21 × 6–10cm, acumen 0.5cm, base rounded then abruptly decurrent on 1–1.8cm petiole, lateral nerves 8–10 pairs; inflorescences axillary, subsessile; flowers 3.5cm wide; corolla white, 5-lobed, tube 8mm; anthers fully exserted, 8mm; fruit globose, orange, 2.5cm when dry. Forest; 1330–1380m.
**Distr.:** Cameroon (Lebialem Highlands) [W Cameroon Uplands].
**IUCN:** CR
**Fosimondi:** <u>Tchiengue, 2212</u> 4/2005; <u>2213</u> 4/2005; <u>2583</u> 2/2006; <u>2584</u> 2/2006; <u>2597</u> 2/2006.
**Local name:** Bush (Tchiengue 2213).

*Argostemma africanum* K.Schum.
Epilithic, erect, succulent, annual herb, 2–14cm, glabrous; leaves anisophyllous, obliquely narrowly elliptic, to 6 × 2cm, erect, obtuse, cuneate, serrate; inflorescences terminal, 1–15-flowered; pedicels glabrous, 1cm; corolla white, divided to the base or almost entire, 5mm diam; fruit a capsule. Wet rocks in forest; 250m.
**Distr.:** SE Nigeria, Cameroon & Rio Muni [lower Guinea].
**IUCN:** NT
**Bechati:** <u>Tchiengue, 2754</u> 9/2006.

*Atractogyne bracteata* (Wernham) Hutch. & Dalziel
Woody climber, to 10m, smelling of liniment, glabrous; leaves ovate, to 20 × 12cm, base obtuse; petiole 4cm; flowers green, cylindric, 10 × 3mm; fruits cylindric, c. 7 × 5cm, red, soft, many-seeded; stipe 4cm. Evergreen forest; 320m.
**Distr.:** Ivory Coast to Gabon [Guineo-Congolian].
**IUCN:** LC
**Bechati:** <u>Tchiengue, 2781</u> 9/2006.
**Uses:** FOOD — Infructescences (Tchiengue 2781).

*Bertiera laxa* Benth.
Shrub, 2m; leaves oblong or elliptic, 23 × 8cm, acumen 1.5cm, base acute; petiole 1cm; stipule lanceolate, 20(–40) × 7mm; flowers c. 20, pendulous in terminal thyrse c. 20cm; branches 5–8, 15mm, 3–4-flowered, pale green; corolla tube 20mm, lobes 2mm; fruit blue. Evergreen forest; 250–1580m.

**Distr.:** SE Nigeria to Gabon [lower Guinea].
**IUCN:** LC
**Bechati:** Tchiengue, 2757 9/2006; 2830 9/2006;
**Fosimondi:** Tchiengue, 2198 4/2005; 2575 2/2006.

### *Chassalia laikomensis* Cheek
Kew Bull. 55(4): 884 (2000).
Shrub, 2–3(–8)m; leaves narrowly elliptic, 4–12 ×
1.5–4cm, acuminate, lateral nerves 7–10 pairs; stipules
4mm, with conspicuous yellow raphides; panicles
terminal, 5 × 5cm, loosely branched; flowers white,
6–10mm long; fruits black, ovoid, 6–9mm long. Forest;
1370–1800m.
**Distr.:** S Nigeria & W Cameroon [W Cameroon
Uplands].
**IUCN:** CR
**Fosimondi:** Tah 292 4/2004; Tchiengue, 1914 4/2004;
2608 2/2006.
Note: Tchiengue 1914 has very long sepals, more
collections needed, Cheek, vii.2005. A similar taxon is
abundant in Kupe-Bakossi & Rumpi Hills. Should it be
shown to be conspecific, the IUCN rating should be
lowered, Cheek, ii.2010.

### *Chassalia simplex* K.Krause (this widely used name
may in future be replaced by *Chassalia subspicata*
K.Schum).
Shrub, 0.5–1m tall; mostly unbranched, glabrous, stems
terete; leaf blade elliptic, c. 18 × 6–9cm including
1.2–2.5cm acumen, base attenuate, lateral nerves 7–11
pairs, uniting 1–2mm from margin, domatia absent,
margin often sinuate; petiole 2–4cm; inflorescence
spike-like at early anthesis, erect, 4–6cm, densely
covered in flowers; corolla s-shaped white-purple, c.
15mm; in late anthesis as corollas fall, spike revealed as
interrupted, with 3 to 6 dense nodes c. 1cm apart;
infructescence red, fruit ovoid, 6mm, branches
accrescent. Along stream edges; 320m.
**Distr.:** SE Nigeria to CAR [lower Guinea].
**IUCN:** NT
**Bechati:** Tchiengue, 2780 9/2006; 2790 9/2006; 2800
9/2006.
Note: *Chassalia simplex* is considered by Lachenaud
(ii.2010) to be a locally common, in some places almost
weedy species (it adapts well to degraded forest) and not
a Cameroon Highland endemic as previously thought.

### *Chassalia* sp. 1 of Bechati
Liana, to 8m, glabrous; stem twisting, hollow; leaves
papery, elliptic to ovate, to 10 × 5cm including the 1cm
slender acumen; base obtuse, lateral nerves 6–7; petiole
1cm; stipules sheathing 2mm, mucra 1 or 2, chaffy;
infructescence light green and dark pink, 3–4cm; fruits
ovoid, 7mm, drying with but one pair of grooves; 2-
seeded. Forest path; 650m.
**Distr.:** Cameroon (Bechati).
**Bechati:** van der Burgt, 886 9/2006.
Note: van der Burgt 886 resembles *Chassalia cristata* but

lacks the densely puberulent inflorescence. *C. cupularis*
occurs to the S, and also climbs, but only reaches 0.5–1m
tall. Flowering material needed, Cheek, i.2010.

### *Coffea montekupensis* Stoffelen
Kew Bull. 52: 989 (1997).
Shrub, 0.6–3m, glabrous; twig bark shiny; leaves
narrowly ovate to obovate, 7–18 × 2.5–5cm, acumen
7–18mm, cuneate, moderately shiny on both faces,
secondary nerves 12 –13 pairs, domatia absent; petiole
3–7mm; stipules triangular, 2–3 × 4–5mm, acute;
inflorescence 1 per axil, 1-flowered; flowers 5-merous;
corolla pink, tube 2.5–4mm, lobes 14–18 × 3–6mm;
fruit ellipsoid, 17–23 × 8–17mm. Forest; 1290m.
**Distr.:** Kupe-Bakossi [W Cameroon Uplands].
**IUCN:** NT
**Fosimondi:** Tchiengue, 2617 2/2006.
Note: Tchiengue 2617 is a very good match for this
taxon, Cheek, vii.2007.

### *Cuviera longiflora* Hiern
Shrub, or tree, to 8m, glabrous; stems with ants; leaves
papery, lanceolate-oblong, c. 27 × 10cm, acuminate,
base subcordate or rounded, nerves 9–10; petiole 1cm;
stipule sheathing, 5mm; flowers 10–20, in axillary
panicles; peduncle 3–8cm; bracts and calyx lobes leafy;
corolla green and white; tube c. 1cm; lobes c. 0.2cm;
fruit ellipsoid, 9 × 3cm, brown, fleshy; pyrenes 5, c. 3 ×
1cm. Evergreen forest; 1800m.
**Distr.:** Cameroon to Angola [lower Guinea].
**IUCN:** NT
**Fosimondi:** Tchiengue, 1905 4/2004.

### *Didymosalpinx abbeokutae* Hiern
Scrambling, spiny, glabrous shrub, to 7m; spines
straight, 0.5–1.5cm, patent, opposite on stem, 0.5cm
above nodes; leaves elliptic, to 17 × 6cm, acumen 1cm,
base obtuse-decurrent, often asymmetric, lateral nerves
c. 8 pairs, domatia raised, circular, margin hairy; petiole
1–1.5cm; stipules 0.5mm, triangular, sheathing;
inflorescence axillary, 1-flowered; pedicels 1–2.5cm;
corolla greenish white, 3.5–8cm; fruit ellipsoid, c. 3 ×
2cm, many-seeded. Forest; 320m.
**Distr.:** Guinea to Cameroon and Congo (Kinshasa)
[Guineo-Congolian].
**IUCN:** NT
**Bechati:** Tchiengue, 2783 9/2006.

### *Galium simense* Fresen.
Straggling herb, clinging by minute hooked hairs on
stems and leaves; leaves 8, in a whorl, linear,
oblanceolate, 10–40 × 2–4mm; flowers solitary,
axillary; pedicels to 1.5cm; fruits globose, 4mm, fleshy,
black. Forest-grassland transition; 1800m.
**Distr.:** Bioko, W Cameroon & Ethiopia [afromontane].
**IUCN:** LC
**Fosimondi:** Tchiengue, 1930 4/2004.

### Geophila afzelii Hiern

Stoloniferous herb, with short vertical shoots, hairy; leaves ovate, 3–6 × 2–5.5cm, subacute, cordate; petioles 3–15cn long, hairy; stipules bifid; inflorescence terminal; peduncle 0.2–5cm; bracts involucral, ovate; bracteoles narrowly lanceolate; flowers white; fruit watery, inflated, red. Forest; 400m.
**Distr.:** Guinea (Conakry) to Congo (Kinshasa) [Guineo-Congolian].
**IUCN:** LC
**Bechati:** van der Burgt, 871 fr., 9/2006.

### Geophila obvallata Didr. subsp. *obvallata*

Creeping herb; stems below ground or at surface; leaves triangular-ovate to ovate-reniform, 3–5 × 2–4cm, apex acute, base cordate, glabrous above, glabrous or slightly pubescent around midnerve beneath; petioles erect, 5–10cm, with two rows of hairs; stipules entire; inflorescence terminal; peduncle erect, 1.5–7cm, glabrous; bracts green, leafy, 7mm, glabrous; flowers white; fruits watery inflated, bright blue or purple, globose, 1cm (live). Farmland and along stream edges 250–350m.
**Distr.:** Guinea (Bissau), Guinea (Conakry), Sierra Leone, Liberia, Ivory Coast, Ghana, Nigeria, Cameroon, Equatorial Guinea, Gabon, Congo (Brazzaville) [upper & lower Guinea].
**IUCN:** LC
**Bechati:** van der Burgt, 851 fr., 9/2006; 855 fr., 9/2006.

### Heinsia crinita (Afzel.) G.Taylor

**Syn.** *Heinsia scandens* Mildbr. nomen.
Wiss. Ergebn. Deutsch. Zentr.-Afr. Exped. 1910-11, 2: 64 (1922).
Shrub, to 7m, appressed grey-puberulous; leaves elliptic, 9 × 3cm, acuminate, 4-nerved, tertiary nerves normal to midrib; petiole 1cm; stipule caducous; flowers terminal, 1 to few; corolla white with yellow hairs in throat, tube c. 2cm; fruit globose, 2cm diam., many-seeded, sepals accrescent, green, 2cm. Evergreen forest; 580m.
**Distr.:** Guinea (Conakry) to Angola [Guineo-Congolian].
**IUCN:** LC
**Bechati:** van der Burgt, 882 9/2006.

### Hekistocarpa minutiflora Hook.f.

Gregarious herb, 1.5m, puberulent; leaves membranous, narrowly elliptic, c. 13 × 3cm, acuminate, acute, 8–10-nerved; petiole 1cm; stipule leafy, triangular, 7mm; inflorescence axillary, tuning-fork shaped, stalk 5mm, branches c. 3cm; corolla white, tube 3mm. Evergreen forest; 350m.
**Distr.:** SE Nigeria & Cameroon [lower Guinea].
**IUCN:** LC
**Bechati:** van der Burgt, 857 9/2006.

### Hymenocoleus hirsutus (Benth.) Robbr.

Bull. Jard. Bot. Nat. Belg. 45: 288 (1975); F.T.E.A. Rubiaceae: 115 (1976).
**Syn.** *Geophila hirsuta* Benth.

Stoloniferous herb with horizontal stems, drying blackish; leaves pubescent above, ovate to ovate-oblong, 2–5 × 1–4cm; stipules shortly divided; inflorescence terminal, without conspicuous leafy bracts, subsessile (rarely pedunculate); flowers white; calyx tube short, lobes much longer than the tube, oblong, hairy; fruit orange, watery inflated; pyrenes with 2 lateral grooves. Forest; 400m.
**Distr.:** Senegal to Congo (Kinshasa) & Tanzania [Guineo-Congolian].
**IUCN:** LC
**Bechati:** van der Burgt, 872 9/2006.

### Hymenocoleus subipecacuanha (K.Schum.) Robbr.

Bull. Jard. Bot. Nat. Belg. 47: 20 (1977).
**Syn.** *Hymenocoleus petitianus* Robbr.
Bull. Jard. Bot. Nat. Belg. 47: 20 (1977).
Stoloniferous, creeping herb; leaves orbicular to ovate, 4.5–6 × 3–6cm, pubescent above and below, or occasionally glabrescent; inflorescence terminal, erect with a cupular involucre divided into triangular lobes for only $^1/_5$–$^1/_3$ of its height; flowers white; calyx tube long, 3–5mm, lobes much shorter than tube; fruit watery inflated, orange; pyrenes with 1 lateral groove. Forest; 590m.
**Distr.:** Nigeria, Cameroon & Congo (Kinshasa) [lower Guinea & Congolian].
**IUCN:** NT
**Bechati:** Tchiengue, 2837 9/2006.
Note: although widespread, apparently rare: only 5 specimens are listed by Robbrecht (1977). However, nearly 43 are now known from Mt Cameroon (Cable & Cheek 1998) & Mt Kupe (Cheek *et al.* 2004).

### Ixora guineensis Benth.

**Syn.** *Ixora breviflora* Hiern
**Syn.** *Ixora talbotii* Wernham
Shrub, 2–3m; leaves coriaceous, oblong, elliptic or obovate-elliptic, c. 14 × 6cm, subacuminate, obtuse, nerves 8, drying yellow-brown below; petiole 1cm; inflorescence sessile, with minute spreading hairs; peduncle c. 2cm; pedicels 3mm; flowers white, corolla tube 25mm. Forest; 1370m.
**Distr.:** Liberia, Nigeria to Congo (Brazzaville) [upper & lower Guinea].
**IUCN:** LC
**Fosimondi:** Tchiengue, 2586 2/2006.

### Keetia hispida (Benth.) Bridson 'setosum'

Kew Bull. 41: 986 (1986).
**Syn.** *Canthium hispidum* Benth.
**Syn.** *Canthium setosum* Hiern
F.T.A. 3: 141 (1877).
Climber, 3m; stems glabrous; leaves drying red below, thickly papery, obovate-elliptic, c. 17 × 8cm, acuminate, unequally truncate-cordate, thinly pilose on both surfaces, domatia inconspicuous, nerves 7–9; petiole 1cm; stipule not seen; fruiting peduncle 0.5cm,

partial-peduncles 1cm; fruit 1.6 × 1.4cm, ridged.
Evergreen forest; 400m.
**Distr.:** Sierra Leone to Congo (Kinshasa) [Guineo-Congolian].
**IUCN:** LC
**Bechati:** van der Burgt, 875 9/2006.
Note: van der Burgt 875 is tentatively placed in this taxon, Cheek, i.2010.

### *Keetia mannii* (Hiern) Bridson s.l.
Kew Bull. 41: 988 (1986).
**Syn.** *Canthium mannii* Hiern
Climber; stems glabrous, square; leaves coriaceous, glossy, elliptic, 10 × 5.5cm, acuminate, obtuse, domatia red-hairy nerves 6–7, petiole 2cm; fruiting peduncle 1.5cm, partial-peduncle 1cm; fruit 1.5 × 2cm.
Evergreen forest; 1400m.
**Distr.:** Sierra Leone to Cameroon, CAR & Sudan [upper & lower Guinea].
**IUCN:** LC
**Fosimondi:** Tah 316 4/2004.

### *Lasianthus batangensis* K.Schum.
Erect shrub, 0.3–2m, pilose; leaves elliptic or slightly obovate, 15–20 × 5–9cm, acutely acuminate, cuneate, midrib pilose above and below; petiole 2–6cm; flowers axillary, several, subsessile, blue or white; corolla 3mm; fruits globose, 8mm, blue. Forest; 250–1580m.
**Distr.:** Sierra Leone to Congo (Kinshasa) [Guineo-Congolian].
**IUCN:** LC
**Bechati:** Tchiengue, 2756 9/2006; 2841 9/2006;
**Fosimondi:** Tchiengue, 2201 4/2005.

### *Lasianthus* sp. 1 of Kupe checklist
The Plants of Kupe, Mwanenguba and the Bakossi Mountains, Cameroon: 375 (2004).
Straggling, or ± climbing herb, to 2–3m, glabrous; stem drying slightly quadrangular; leaves elliptic, 8 × 3cm, glabrous below, lateral nerves 8–10 pairs; petiole 1.5cm; corolla white; fruit 3–4mm. Forest; 1400m.
**Distr.:** Cameroon, Gabon, Congo (Brazzaville) [lower Guinea].
**Fosimondi:** Tchiengue, 1949 4/2004.

### *Massularia acuminata* (G.Don) Bullock ex Hoyle
Shrub, 4m; stems glabrous; leaves obovate-elliptic, c. 30 × 10cm, acuminate, subcordate, nerves 15–20, subsessile; stipule semi-circular, c. 4 × 1.5cm; flowers 5–10, axillary, panicle branched, 5cm; corolla pink and white; tube 12cm; lobes 8mm; fruit spherical-rostrate, 8 × 6cm, green, fleshy, few-seeded. Evergreen forest; 250m.
**Distr.:** Guinea (Conakry) to Congo (Kinshasa) [Guineo-Congolian].
**IUCN:** LC
**Bechati:** Tchiengue, 2759 9/2006.
Note: Tchiengue 2759 is a field determination, material not seen at Kew, Harvey, ii.2010.

### *Mussaenda erythrophylla* Schum. & Thonn.
Climber, densely puberulent; leaves elliptic, to 1.5 × 7cm, acuminate, obtuse, lateral nerves 9–13 pairs; petiole to 2cm; flowers 10–20; calyx with one leaf-sized, bract-like, red lobe; corolla cream to orange; tube 2cm; lobes 0.5cm. Evergreen forest; 1486m.
**Distr.:** Guinea (Conakry) to S Africa [tropical Africa].
**IUCN:** LC
**Fosimondi:** Tchiengue, 2206 4/2005.

### *Mussaenda tenuiflora* Benth.
Climber, pubescent, resembling *M. erythrophylla*, but bract-like calyx lobes white, not red. Evergreen forest; 320–1400m.
**Distr.:** Guinea (Conakry) to Congo (Kinshasa) [Guineo-Congolian].
**IUCN:** LC
**Bechati:** Tchiengue, 2798 9/2006; **Fosimondi:** Tah 302 4/2004; Tchiengue, 2632 2/2006.
Note: Tchiengue 2798 and 2632 are tentatively placed in this taxon (determined as *M.* cf. *tenuiflora*, Tchiengue, vii.2007 and Cheek, i.2010 respectively).

### *Mussaenda* sp. 1 of Fosimondi
Subshrub, to 0.6m; stem 10cm, with two foliaceous nodes at apex; leaf blades ovate, 9 × 7cm, acumen with mucron, base obtuse, lateral nerves 5 pairs, upper and lower surface pink to purplish when live, with scattered appressed white bristle hairs; petiole 1cm; flowers terminal, inflorescence axis and bracts absent; pedicels 10mm, glabrous; ovary 4mm; calyx lobes 12mm, linear, all equal; corolla bud 15mm, appressed hairy; flowers yellow when live. Submontane forest; 1380m.
**Distr.:** Cameroon (Fosimondi) [W Cameroon Uplands].
**Fosimondi:** Tchiengue, 2226 4/2005.
Note: very similar and possibly conspecific with *Mussaenda epiphytica* of Bakossi but differing in the calyx lobes being all equal. Although the field notes of Tchiengue 2226 do not indicate that epiphytic, the material may have fallen from a tree. Careful comparison is needed and better material, Cheek, i.2010.

### *Otomeria cameronica* (Bremek.) Hepper
Herb; stems 0.3–1m long, usually prostrate; leaves ovate-elliptic or ovate-lanceolate, 2–8 × 0.4–4cm; inflorescence terminal; flowers white; enlarged calyx lobe ovate; corolla-tube 3–5mm long; fruit a capsule, ovoid. Forest and grassland.
**Distr.:** Sierra Leone to Bioko & Cameroon [Guineo-Congolian (montane)].
**IUCN:** LC
**Fosimondi:** Tchiengue, 2259 4/2005.

### *Oxyanthus formosus* Hook.f. ex Planch.
Shrub or tree, 3–10m; leaves coriaceous, oblong to 25 × 12cm, short-acuminate, unequally obtuse; petiole 1cm; stipule lanceolate, 2–3 × 1–1.5cm; corolla tube 12 × 0.1cm; lobes 2.5 × 0.1cm; fruit ellipsoid, 5 × 2.5cm; pedicel 2.5cm. Evergreen forest; 1290–1370m.

Distr.: Mali to Uganda [Guineo-Congolian].
IUCN: LC
Fosimondi: Tchiengue, 2211 4/2005; 2621 2/2006.
Local name: Bush Coffee (Pidgin) (Tchiengue 2211).

*Oxyanthus gracilis* Hiern
Shrub, 0.6–2m; leaves elliptic, c. 15 × 6cm, acuminate,
acute; petiole 1cm; stipule ovate, 6–10 × 4–6mm;
corolla tube 5.2–7cm, lobes 1–1.2cm; fruit spherical,
2.5cm, pedicel 6mm. Evergreen forest; 530m.
Distr.: S Nigeria, Bioko to Congo (Kinshasa) [lower
Guinea & Congolian].
IUCN: NT
Bechati: Tchiengue, 2831 9/2006.

*Pauridiantha floribunda* (K.Schum. & K.Krause)
Bremek.
Tree, 6–12m; stem lacking ant cavities, glabrous; leaves
lanceolate-oblong, c. 15–20 × 5–6cm, acuminate, base
rounded or obtuse, drying pink below, nerves c. 20
pairs; petiole 1.5–2cm; stipule leafy, ovate, c. 1.5 ×
0.8cm, acuminate; flowers c. 70 in erect, flat-topped
inflorescences, c. 10cm, white; peduncle c. 10cm; fruit
green, 8mm. Evergreen forest; 1400–1800m.
Distr.: SE Nigeria to São Tomé & Gabon [lower
Guinea].
IUCN: NT
Fosimondi: Tah 318 4/2004; Tchiengue, 1936 4/2004.

*Pauridiantha paucinervis* (Hiern) Bremek.
Shrub, 2–4m; stems puberulent; leaves elliptic-oblong,
c. 10 × 3cm, base acute, nerves 8, domatia usually
absent; petiole 7mm; stipule subulate, 7 × 1mm;
flowers 5–10, on 1–2 peduncles to 1.5cm, white;
corolla tube 3mm; fruit 5mm, red or black. Evergreen
forest; 1800m.
Distr.: widespread in highlands of tropical Africa &
Madagascar [afromontane & submontane].
IUCN: NT
Fosimondi: Tchiengue, 1913 4/2004.
Note: despite its broad distribution its habitat has seen
ongoing loss across its range, Cheek, ii.2010.

*Pavetta bidentata* Hiern var. *bidentata*
Shrub, 2(–4)m, glabrous; floriferous twigs to 30cm;
leaves thinly leathery, mostly narrowly oblong-elliptic,
to 25 × 6cm, acute, cuneate, midrib drying orange,
lateral nerves 15 pairs, domatia pits elongate, hairy,
nodules elongate along midrib, rare in blade, tertiary
venation just visible; petiole to 3.5cm; inflorescence
4cm across, subglabrous; corolla white, tube 4–10 ×
1–2mm, lobes 5–10mm; fruit globose, pink or white
with green stripes, then black; seeds 1–2, concave.
Forest; 400m.
Distr.: SE Nigeria to Congo (Kinshasa) [Guineo-
Congolian].
IUCN: LC
Bechati: van der Burgt, 868 9/2006.

*Pavetta brachycalyx* Hiern
Shrub, 3–6m, glabrous; floriferous twigs 20cm; leaves
papery, drying green, elliptic to 20 × 9cm, acumen
0.5–1cm, acute, lateral nerves 9 pairs, tertiary nerves
inconspicuous, domatia small pit-like, bacterial nodules
rod-like, 2–3mm, sparse; petioles 3cm; panicles 10cm
across; calyx lobes 0.5mm; corolla white, tube 6mm,
lobe 6mm. Forest; 400m.
Distr.: Cameroon [W Cameroon Uplands].
IUCN: EN
Bechati: van der Burgt, 874 9/2006.
Note: once thought restricted to Mt Cameroon. Easily
confused with *P. gabonica* (q.v.).

*Pavetta calothyrsa* Bremek.
Ann. Missouri Bot. Gard. 83: 107 (1996).
Shrub, to 3m, glabrous; leaves opposite, glossy above,
elliptic, 15–25 × 7–10cm, acumen indistinct, base
acute-decurrent, lateral nerves 7 pairs, domatia absent;
petioles 3–4(–6)cm; stipule sheath 4mm, awn 1.5mm;
inflorescence terminal, c. 6 × 7cm, white puberulent,
dense-flowered; flower bud 7mm long; fruit globose,
8mm, glossy; calyx lobes square, 4, erect, as long as
tube. Edge of gallery forest; 1370m.
Distr.: Cameroon, Gabon & Congo (Kinshasa) [lower
Guinea & Congolian].
IUCN: LC
Fosimondi: Tchiengue, 2595 2/2006.
Note: we follow Manning (Ann. Miss. Bot. Gard. 83:
107 (1996)) in restoring *P. calothyrsa* from synonymy
under *P. nitidula*. This species has been reported from
Gabon and Congo (Kinshasa) as well as Cameroon
(Cheek, vi.2009).

*Pavetta gabonica* Bremek.
Ann. Missouri Bot. Gard. 83: 113 (1996).
Shrub, similar to *P. brachycalyx*, but flowers yellow,
smaller (corolla tube 2–4mm, not 3–5mm, lobe 3–5mm,
not 4–6mm); calyx lobes 0.75mm; inflorescence
1–5.5cm across (not 1–10cm); fruit orange. Stream
banks in forest; 320m.
Distr.: Cameroon & Gabon [lower Guinea].
IUCN: NT
Bechati: Tchiengue, 2792 9/2006.
Note: Tchiengue 2792 is tentatively placed in this
taxon, Tchiengue, vii.2007.

*Pavetta hookeriana* Hiern var. *hookeriana*
Shrub, 2–3m, subglabrous; floriferous twigs 15cm;
leaves papery, elliptic to 13 × 6cm, acumen 0–1cm,
cuneate, lateral nerves 10 pairs, domatia arched, hairy,
nodules not seen, tertiary venation inconspicuous; petiole
2cm; inflorescence to 10cm across; flowers to 100; calyx
lobes rotund, 2mm; corolla white; tube 2–5mm; lobes
4–8mm. Forest-grassland transition; 1700–1800m.
Distr.: Bioko & W Cameroon [W Cameroon Uplands].
IUCN: VU
Fosimondi: Tah 289 4/2004; Tchiengue, 1922 4/2004.

Note: both Tah 289 & Tchiengue 1922 are atypical forms of this taxon. Differs from *P. hookeriana* in leaves drying brown, the persistent bracts and the domatia.

**Pavetta owariensis** P.Beauv. var. **owariensis**
Shrub, c. 3m, glabrous; leaves elliptic, c. 25 × 10cm, acumen 1cm, base cuneate, lateral nerves 5–6 pairs, not looping, quaternary nerves visible to naked eye, domatia hairs present, bacterial nodules rectangular, black, from thickened nervelets; petiole 3–5cm; stipule sheath 2–3mm, limb reduced to arista; inflorescence on stout lateral 'twiglets' c. 60cm long; leaf-like bracts c. 10cm; infructescence 10–15cm wide; fruits white-brown globose, 1cm; calyx remains un-lobed. Forest; 590m.
**Distr.:** Sierra Leone to Cameroon [upper & lower Guinea].
**IUCN:** LC
**Bechati:** Tchiengue, 2818 9/2006.
Note: this variety is rare in Cameroon and recognised by having tertiary nerves visible to the naked eye and hairy domatia, Cheek, i.2010.

**Pentaloncha sp. nov. of Kupe Bakossi checklist**
The Plants of Kupe, Mwanenguba and the Bakossi Mountains, Cameroon: 381 (2004).
Subshrub, 0.3–0.5m; leaves membranous, elliptic, 7 × 2.5cm, subacuminate, acute, lateral nerves 7–10 pairs; petiole 1–2cm; stipule aristate, 5mm; inflorescences axillary, fascicles 5–10-flowered, sessile; corolla 2–3mm, purple; fruit globose, 2mm. Forest; 1310m.
**Distr.:** Cameroon (Bakossi Mts & Fosimondi) [Cameroon Endemic].
**Fosimondi:** Tchiengue, 2616 2/2006.
Note: Tchiengue 2616 compared to the description in the Kupe checklist, Cheek, i.2010.

**Pentas schimperiana** (A.Rich.) Vatke subsp. **occidentalis** (Hook.f.) Verdc.
Herb, or subshrub, 1.5m; leaves ovate-elliptic, 9 × 4cm, pubescent below; petiole 1cm; flowers yellow white; corolla tube 17mm long. Grassland, forest edge; 1800m.
**Distr.:** Cameroon, Bioko, São Tomé & Ituri (Congo (Kinshasa)) [Guineo-Congolian (montane)].
**IUCN:** LC
**Fosimondi:** van der Waarde, 20 10/2004.
Note: for more information about this taxon, see Blumea 53: 566-568 (2008).

**Poecilocalyx schumannii** Bremek.
Fl. Gabon 12: 230 (1966).
Shrub, 3.5m; stems villose; leaves papery, narrowly elliptic, 15 × 4cm, acuminate, rounded-obtuse, pubescent below, nerves 8–9; petiole 5mm; stipule palmately 5-lobed, villose; flower greenish brown, 5mm diam.; fruit hemispherical, 1cm; calyx lobes leafy. Evergreen forest; 250–320m.
**Distr.:** Cameroon, Equatorial Guinea (Rio Muni) & Gabon [lower Guinea].
**IUCN:** LC

**Bechati:** van der Burgt, 854 9/2006; Tchiengue, 2782 9/2006.

**Psychotria babatwoensis** Cheek
Kew Bull. 63: 414 (2008).
**Syn. Psychotria sp. A of Bali Ngemba checklist**
The Plants of Bali Ngemba Forest Reserve, Cameroon: 123 (2004).
Shrub, 2m, glabrous; leaves elliptic, thickly papery, drying grey-green above, pale below, bacterial nodules conspicuous, punctate, in clusters, 10 × 6cm, acumen 1cm, acute-decurrent, lateral nerves 9 pairs, brown; petiole 1.5cm; stipule ovate, 6mm, apical third bifid; panicle 2 × 2cm; peduncle 1cm, 10–20-flowered; corolla 3mm, white; infructescence 5cm; peduncle 2-winged; fruits 10–20, ellipsoid, red, 14 × 6mm, smooth. Submontane forest with *Pterygota mildbraedii*; 1355m.
**Distr.:** Cameroon (Bali Ngemba FR, Baba 2 community forest & Fosimondi) [W Cameroon Uplands].
**IUCN:** EN
**Fosimondi:** Tchiengue, 2609 2/2006.

**Psychotria camptopus** Verdc.
Kew Bull. 30: 259 (1975).
**Syn. Cephaelis mannii** (Hook.f.) Hiern
Tree, 3–5m, glabrous; leaves leathery, obovate, 26 × 13cm, acumen obtuse, 1cm, acute, lateral nerves 14 pairs; petiole 6cm; stipules 2 × 1.5cm; peduncle 1–3m, pendulous, red; flowers white outside, yellow within, 1cm, 10–20, enveloped in red fleshy bracts; involucre 3 × 5cm, glabrous. Forest; 1800m.
**Distr.:** SE Nigeria, Bioko, SW Cameroon [W Cameroon Highlands].
**IUCN:** NT
**Fosimondi:** Tchiengue, 1909 4/2004.

**Psychotria globosa** Hiern var. **globosa**
Monopodial herb, to 0.6m, sparsely puberulent, glabrescent; leaves drying black, elliptic or obovate, 13 × 6cm, subacuminate, acute, lateral nerves 8–10 pairs; petiole 1cm; stipule oblong-elliptic, 12mm; panicle capitate, dense, flat-topped, 3cm; flowers white, 9mm; fruit ovoid, 1cm, red, smooth. Forest; 320m.
**Distr.:** Nigeria, Cameroon, Equatorial Guinea (Bioko & Rio Muni) [lower Guinea].
**IUCN:** NT
**Bechati:** Tchiengue, 2785 9/2006; 2786 9/2006.

**Psychotria lucens** Hiern var. **lucens**
F.T.A. 3: 211 (1877).
**Syn. Psychotria calceata** *sensu* Cheek in The Plants of Kupe, Mwanenguba and the Bakossi Mountains, Cameroon: 383 (2004).
Shrub, to 0.8m, glabrous; leaves sometimes dimorphic on stem: elliptic, with base obtuse-acute and also broadly obovate, base abruptly decurrent; blade 15–20 × 8–14cm, acuminate, lateral nerves 12–14 pairs; petiole 2–3cm; stipule 9mm, deeply bifid; inflorescence 30cm, diffuse; peduncle c. 10cm, lowest node 3 or 4-

branched, branches 9cm; fruit red, ellipsoid, 9mm. Forest, stream banks; 400m.
**Distr.:** Nigeria, Cameroon, Equatorial Guinea (Bioko), São Tomé & Príncipe [lower Guinea & Gulf of Guinea islands].
**IUCN: NT**
**Bechati:** van der Burgt, 866 9/2006.

## *Psychotria martinetugei* Cheek
Kew Bull. 57: 375–387 (2002).
**Syn.** *Psychotria malchairei sensu* Hepper
Shrub, 0.2–4m, glabrous, fistular; leaves papery, elliptic, 12–18 × 4–6cm, acumen 0.2–1.2cm, cuneate, secondary nerves 10–12 pairs; petiole 2–4cm; stipules caducous, 4–6mm, acumen 1mm, entire or bifurcate; panicles 4–6 × 5–9cm, 30–60-flowered; peduncle 3mm; flowers white, 2.5mm; corolla 5-lobed; fruit ellipsoid, 1cm, on dilated pedicel. Forest; 1350–1370m.
**Distr.:** SE Nigeria, Bioko, Cameroon [lower Guinea].
**IUCN: NT**
**Fosimondi:** Tchiengue, 2218 4/2005; 2607 2/2006.

## *Psychotria peduncularis* (Salisb.) Steyerm. var. *hypsophila* (K.Schum. & K.Krause) Verdc.
Kew Bull. 30: 257 (1975).
**Syn.** *Cephaelis peduncularis* Salisb. var. *hypsophila* (K.Schum. & K.Krause) Hepper
Shrub, 1–5m, glabrous; leaves elliptic, to 15 × 8cm, acumen 0.5cm, acute, lateral nerves 12–15 pairs; petiole 2–3cm; stipule translucent, bifurcate, 1 × 0.8cm; inflorescence capitate; peduncle 2–4cm, nodding in accrescence, glabrous; involucral bracts fleshy; flowers 10–15, white, 5mm; infructescence umbellate; bracts fallen; pedicels white, 1.5cm; berries blue, 7mm. Forest; 1700m.
**Distr.:** Bioko & Cameroon [W Cameroon Uplands].
**IUCN: LC**
**Fosimondi:** Tchiengue, 1942 4/2004.
Note: the status of this and other varieties of *P. peduncularis* needs further study, Lachenaud, ii.2010.

## *Psychotria peduncularis* (Salisb.) Steyerm. var. *suaveolens* (Schweinf. ex Hiern) Verdc.
Kew Bull. 30: 257 (1975).
**Syn.** *Cephaelis peduncularis* Salisb. var. *suaveolens* (Schweinf. ex Hiern) Hepper
Shrub, 1.5cm, glabrous; leaves elliptic, c. 13 × 6cm, acumen 1cm, base acute, lateral nerves 12 pairs, impressed above, patent brown puberulent below, domatia absent; petiole 2cm; inflorescence terminal; peduncle 3cm, erect, puberulent in 2 longitudinal lines; capitula 2cm wide, enveloped in bracts; flowers white, densely packed, c. 2mm wide. Disturbed edge of forest; 320m.
**Distr.:** Cameroon, Congo (Kinshasa), Ghana, Guinea, Uganda, Kenya, Tanzania, Malawi, Sudan, Zambia [tropical Africa].
**IUCN: LC**
**Bechati:** Tchiengue, 2805 9/2006.

Note: the status of this and other varieties of *P. peduncularis* needs further study, Lachenaud, ii.2010.

## *Psychotria succulenta* (Hiern) Petit
Shrub, 1–3m, drying dark brown, matt; leaves leathery, elliptic-oblong, 15 × 7cm, acumen 0.5cm, acute-obtuse, lateral nerves 12 pairs; petiole 1cm; stipule broadly elliptic, 1.5cm, entire; inflorescence paniculate (sometimes shortly so), 3 × 3cm; peduncle 7cm; flowers white, 3mm; fruit ovoid, 5mm. Forest.
**Distr.:** Ghana and Nigeria to Zambia [Sudano-Zambesian].
**IUCN: LC**
**Fosimondi:** Tchiengue, 2265 4/2005.

## *Psychotria* sp. 9 of Kupe-Bakossi
The Plants of Kupe, Mwanenguba and the Bakossi Mountains, Cameroon: 388 (2004).
Shrub, 2–5m, minutely puberulent, glabrescent; leaves oblanceolate, to 12 × 4cm, with lateral nerves 9–10 pairs; petiole to 2cm; stipule 7mm; inflorescence loosely capitate with two axillary inflorescences subequal to the main inflorescence; flower cluster to 1.5 × 2cm, to 20-flowered; peduncle to 1.5cm, glabrous; flowers white, 3mm; fruit globose, 1cm, red. Forest; 1800m.
**Distr.:** Cameroon (Bakossi & Fosimondi) [W Cameroon Uplands].
**Fosimondi:** Tchiengue, 1921 4/2004.

## *Psychotria* sp. B of Bali Ngemba checklist
The Plants of Bali Ngemba Forest Reserve, Cameroon: 123 (2004).
Shrub, 0.5–1.5m; stems, stipules and inflorescences densely brown puberulent; leaves papery, drying dark green above, lacking bacterial nodules, elliptic, 13 × 6cm, acute to subacuminate, base obtuse, lateral nerves 9–10 pairs, arched, white above, lacking domatia; petioles 1.5cm; stipules to 1.4 × 0.7cm, upper quarter bifid; inflorescence densely capitate, 1.5 × 1.2cm; peduncle 0.5cm; corolla tube 2mm, white; infructescence erect, to 4.5 × 4cm; peduncle 1.5cm; fruits numerous, subglobose, 5mm. Forest; 1300m.
**Distr.:** known only from Bali Ngemba FR & Fosimondi [Cameroon endemic].
**Fosimondi:** Tah 309 4/2004.

## *Psydrax kraussioides* (Hiern) Bridson
F.T.E.A. Rubiaceae: 907 (1991); Fl. Zamb. 5(2): 362 (1998).
**Syn.** *Canthium kraussioides* Hiern
**Syn.** *Canthium henriquezianum sensu* Hepper
Woody climber; leaves elliptic, shiny coriaceous, 9–13 × 4–6cm; stipules with a keeled lobe; flowers yellow-green; fruit 2-lobed (each almost spherical). Forest patches, roadsides and villages.
**Distr.:** Guinea to Angola [Guineo-Congolian].
**IUCN: LC**
**Fosimondi:** Tchiengue, 2256 4/2005.

**Uses:** FOOD — Seeds: seed edible when fruit is ripe (Tchiengue 2256). **Local name:** Lepah (Banwa) (Tchiengue 2256).

### *Rothmannia whitfieldii* (Lindl.) Dandy
Tree, 12m; leaves leathery, glabrous, elliptic, drying brown below, c. 24 × 8cm, acumen 1.5cm, acute-obtuse, lateral nerves 8 pairs, venation conspicuous below; petiole 1.5cm. calyx lobes 15–66mm; corolla tube long velutinous, 3–17cm; fruit subglobose, 7cm, 10-ridged. Forest; 590m.
**Distr.:** tropical Africa.
**IUCN:** LC
**Bechati:** Tchiengue, 2819 9/2006.

### *Rutidea decorticata* Hiern
Climber, glabrous or glabrescent; epidermis of stem and petiole exfoliating; leaves coriaceous, elliptic, to 15 × 8cm, base obtuse, nerves 4–8, impressed and white above, finely reticulate below, domatia inconspicuous; petiole 1.5cm; stipule triangular, aristate; inflorescence with numerous branches; corolla white, tube 3mm; fruit orange. Evergreen forest; 1400–1500m.
**Distr.:** Nigeria to Congo (Kinshasa) [lower Guinea & Congolian].
**IUCN:** LC
**Fosimondi:** Tah 305 4/2004; 307 4/2004; Tchiengue, 2267 4/2005.

### *Rutidea olenotricha* Hiern
Climber, densely and shortly brown pubescent; leaves elliptic, oblong or oblanceolate, to 15 × 8cm, shortly acuminate, obtuse to rounded, nerves 7–8, domatia large, bright brown hairy, extending along secondary nerves; petiole 10mm; inflorescence with numerous branches, corolla yellow, tube 5mm; fruit yellow, 6mm. Evergreen forest; 550m.
**Distr.:** Sierra Leone to Congo (Kinshasa) [Guineo-Congolian].
**IUCN:** LC
**Bechati:** Tchiengue, 2829 9/2006.

### *Rutidea smithii* Hiern subsp. *smithii*
Climber, grey puberulent or glabrescent; leaves papery, drying matt black-brown above, grey-brown below, elliptic or elliptic-obovate, 10–17 × 4–8cm, shortly acuminate, acute, nerves 7–9, with bright white hairy domatia extending to tertiary nerve junctions; petiole 1–2.5cm; stipule awn 8mm; inflorescence with numerous branches; corolla white, tube 3mm; fruit green, 6mm. Evergreen forest; 1600m.
**Distr.:** Sierra Leone to Kenya [tropical Africa].
**IUCN:** LC
**Fosimondi:** Tchiengue, 2565 2/2006.

### *Sabicea calycina* Benth.
Climber, puberulous; leaves membranous, oblong, c. 9 × 4cm, acuminate, cordate, whitish-green below, nerves 9; petiole to 3.5cm; flowers 10–15; peduncle c. 6cm; bracts ovate, 1.2cm; calyx lobes elliptic, purple, 1.2cm, corolla white, tube 2cm. Forest; 1300–1400m.
**Distr.:** Sierra Leone to Congo (Kinshasa) [Guineo-Congolian].
**IUCN:** LC
**Fosimondi:** Tah 306 4/2004; Tchiengue, 1956 4/2004.

### *Sabicea venosa* Benth.
Climber, appressed white pubescent; leaves thinly papery, elliptic, to 9 × 3.5cm, acuminate, rounded to obtuse, nerves 16; petiole 1cm; flowers 5–10; peduncle 0.5cm; branches 0.5cm; bracts inconspicuous; calyx lobes 3mm; corolla white, tube 5mm; fruit 1cm, globose, white. Evergreen forest; 1500m.
**Distr.:** Senegal to Congo (Kinshasa) [Guineo-Congolian].
**IUCN:** LC
**Fosimondi:** Tchiengue, 2271 4/2005.
Note: *S. venosa* is a complex aggregate in need of further study, Lachenaud, ii.2010.

### *Sabicea xanthotricha* Wernham
Shrub, 3m, cauliflorous, pubescent; leaves membranous, elliptic, c. 30 × 17cm, subacuminate, rounded-decurrent, nerves 20; petiole 10cm; flowers numerous, sessile; calyx lobes filiform, c. 1cm; corolla white, tube c. 1cm. Forest; 1400m.
**Distr.:** SE Nigeria & Cameroon [lower Guinea].
**IUCN:** EN
**Fosimondi:** Tchiengue, 1954 4/2004.

### *Sherbournia zenkeri* Hua
Climber; leaves 9–14 × 3.3–7cm, lateral nerves 10–12 pairs, densely appressed puberulent below; calyx purple, tube 3mm, lobes 12–15mm; corolla white, tube 26–32mm, lobes 10mm; fruits cylindric, 3.5 × 1.5cm, strongly ribbed. Forest; 530–1370m.
**Distr.:** Nigeria to Cabinda [lower Guinea].
**IUCN:** NT
**Bechati:** Tchiengue, 2814b 9/2006; 2842 9/2006;
**Fosimondi:** Tchiengue, 2222 4/2005.

### *Spermacoce princeae* (K.Schum.) Verdc. var. *princeae*
F.T.E.A. Rubiaceae: 362 (1976).
**Syn.** *Borreria princeae* K.Schum. var. *princeae*
Creeping sometimes scandent herb, to about 60cm; stems 4-angled, 3mm wide, glabrous; internodes 4–12cm; leaves ovate to ovate-lanceolate, 2–5 × 1–2cm, lateral nerves 5 on each side, deeply impressed above, prominent below, puberulent; petiole 3–5mm; stipule cup 6 × 6mm, apex truncate, with 10 equally spaced bristles 6–10mm long; inflorescence axillary; flowers sessile; 4-lobed calyx; corolla white, 4-lobed, 8mm wide; fruit a capsule, oblong-ellipsoid, 5–6mm long. Forest, forest-grassland transition; 1800m.
**Distr.:** Cameroon to Tanzania [Guineo-Congolian (montane)].
**IUCN:** LC
**Fosimondi:** Tchiengue, 2283 4/2005.

### *Tarenna eketensis* Wernham

Climber, densely brown-puberulent on stem, lower surface of leaves and inflorescence; leaves elliptic or obovate, 9(–14) × 3.5(–6.5)cm, acumen broad, rounded, base acute, asymmetric, nerves 5–7, domatia hairy, extending to secondary branches; petiole 7mm; flowers c. 30; peduncles 15mm; pedicels 2–3mm, white; corolla tube 5mm, lobes 3mm; fruit globose, 8mm, arachnoid hairy. Evergreen forest; 540m.
**Distr.:** Liberia to Congo (Kinshasa) [Guineo-Congolian].
**IUCN:** LC
**Bechati:** Tchiengue, 2828 9/2006.

### *Tarenna fusco-flava* (K.Schum.) N.Hallé

Adansonia (sér. 2) 7: 506 (1967); Kew Bull. 34: 379 (1979); F.T.E.A. Rubiaceae: 588 (1988).
**Syn.** *Tarenna flavo-fusca* (K.Schum.) S.Moore
Climber, 7m, glabrous; leaves elliptic, c. 6 × 3cm, spathulate-acuminate, unequally acute, nerves 2–3, hairy pocket domatia; petiole 4mm; stipule ovate-acuminate, 4 × 2mm; flowers c. 3 in subterminal axils; pedicels 3cm; corolla tube pink, 1.5cm, lobes 0.8cm; fruit globose, 8mm. Evergreen forest; 350m.
**Distr.:** Liberia to Uganda [Guineo-Congolian].
**IUCN:** LC
**Bechati:** Tchiengue, 2763 9/2006.

### *Tarenna lasiorachis* (K.Schum. & K.Krause) Bremek.

Shrub, 1–3m; stems and lower surface fo leaves densely appressed puberulent; leaves elliptic to 16 × 6cm, acumen 1cm, acute, lateral nerves c. 10 pairs; petiole 5mm; stipule glossy black, triangular, 2mm; inflorescence terminal, dense, 5cm wide; calyx lobes 2.5mm; corolla white tube 10mm, 5 lobes each 6mm; fruit globose, 7mm. Forest; 530m.
**Distr.:** Cameroon, Gabon, Congo (Brazzaville) [lower Guinea].
**IUCN:** NT
**Bechati:** Tchiengue, 2814b 9/2006.

### *Tarenna vignei* Hutch. & Dalziel var. *subglabra* Keay

Shrub, 2m, stem and leaves glabrous; leaf-blade narrow elliptic, to 18 × 5cm, acumen 2cm, curved, base asymmetric, cuneate, lateral nerves 4–7 pairs, domatia hairy elliptic craters; petiole 10–25mm; stipule glossy, black, 2mm; inforescence terminal, 4 × 8cm, c. 12-flowered; pedicels 5mm, puberulous; calyx glabrous, lobes 1mm; corolla white, tube 9mm, lobes 5. Forest; 1450m.
**Distr.:** Sierra Leone to Cameroon [Guineo-Congolian].
**IUCN:** NT
**Fosimondi:** Tchiengue, 2184 4/2005.

### *Tricalysia discolor* Brenan

**Syn.** *Tricalysia mildbraedii* Keay
Tree, 8–10m, glabrous; leaves drying grey-brown, papery, elliptic or elliptic-oblong, acuminate, acute, nerves 7, tertiary nervation scalariform; petiole 1.5cm; inflorescences several, subsessile; bracts cup-shaped; flowers numerous; bracteoles alternate on pedicel 5mm; fruit elliptic, 5mm; calyx lobes triangular. Forest; 590m.
**Distr.:** Liberia to Cameroon [upper & lower Guinea].
**IUCN:** LC
**Bechati:** Tchiengue, 2817 9/2006.

### *Tricalysia* sp. B aff. *ferorum* Robbr.

Tree, 7–13m, densely white-appressed-hairy; leaves thinly coriaceous, oblong-acuminate, acumen abrupt, 7mm, base acute-decurrent, nerves 9; petiole 1.2cm; inflorescence on naked stem, 3-flowered, both bracts and bracteoles united, disc-like; calyx 5-lobed; corolla tube 5mm; lobes 5; fruit green, globose, 1cm, disc 5mm diam. Forest; 1800m.
**Distr.:** Mt Kupe & Bali Ngemba [Cameroon endemic].
**Fosimondi:** Tchiengue, 1915 4/2004.
Note: determined as *Tricalysia* not matched, near sp. B aff. *ferorum* of Mt Kupe & Bali Ngemba. Tchiengue 1915 differs in domatia; smaller leaves and longer pedicels, Cheek, iii.2005.

### *Trichostachys petiolata* Hiern

F.T.A. 3: 227 (1877).
**Syn. Trichostachys sp. 1 of Kupe Bakossi**
The Plants of Kupe, Mwanenguba and the Bakossi Mountains, Cameroon: 395 (2004).
Prostrate herb, pubescent stems, petioles, blade veinlets below; flowering stems erect, c. 10cm, 2–4-nodes; leaf-blades in rosettes, obovate, elliptic, c. 10 × 6cm, apex and base rounded, lateral nerves 10–11 pairs, tertiary nerves conspicuous, subscalariform; petiole 1cm; stipule ovate-triangular, 7 × 4mm; inflorescence erect, spike 2.5–4cm; peduncle 2cm, white hairy; corolla white; tube 3mm; lobes 1mm. Forest and farmland; 650m.
**Distr.:** Cameroon [Cameroon Endemic].
**IUCN:** EN
**Bechati:** van der Burgt, 885 fl., 9/2006.
Note: first collected 1861, Cameroon River (Mann 721), recollected in Bakossi in 1998 (Etuge 4296) (needs confirmation), Cheek, i.2010.

### *Virectaria procumbens* (Sm.) Bremek

Herb, 20cm, straggling; stems with 2 lines of pubescence; leaves ovate-oblong to subspathulate, 1–6 × 0.5–3.5cm; stipules triangular, 3mm, entire or cleft; inflorescence terminal, few flowered; flowers white; calyx lobes spathulate; corolla 1cm long; fruit a capsule, with a raised disc 1mm long. Forest; 590m.
**Distr.:** Guinea to Congo (Kinshasa) [Guineo-Congolian].
**IUCN:** LC
**Bechati:** Tchiengue, 2822 9/2006.

# RUTACEAE

Y.B. Harvey (K) & M. Cheek (K)

Fl. Cameroun 1 (1963).

## Clausena anisata (Willd.) Hook.f. ex Benth.

Shrub or tree, 3–8m, non-spiny, puberulent, strongly aromatic; leaves imparipinnate, 15cm, 4–9-jugate; leaflets alternate, lanceolate-oblique, c. 6 × 2.5cm, acuminate, obtuse, lateral nerves c. 10 pairs; petiolules 1mm; panicle c. 12cm, slender; flowers white, 5mm; fruit indehiscent. Forest; 1500m.

**Distr.:** Guinea (Conakry) to Malawi [tropical Africa].
**IUCN:** LC
**Fosimondi:** Tchiengue, 2266 4/2005.

## Zanthoxylum rubescens Hook.f.

F.T.E.A. Rutaceae: 44 (1982); Hawthorne W., F.G.F.T. Ghana: 169 (1990).
**Syn.** *Fagara rubescens* (Hook.f.) Engl.
Shrub or tree, 2.5–20m; leaves to c. 0.6m, 4(–6)-jugate, rachis smooth; leaflets drying pale green below, oblong, to 18 × 7cm, lateral nerves c. 10; petiolule 12mm; inflorescence a terminal panicle, 20cm; flowers numerous, 5mm, greenish or whitish; fruit globose, pink, 1cm, dehiscent, seed glossy. Farmbush; 1700m.
**Distr.:** Guinea (Bissau) to Angola [Guineo-Congolian].
**IUCN:** LC
**Fosimondi:** Tah 286 4/2004.

# SAPINDACEAE

Y.B. Harvey (K), M. Cheek (K) & B. Tchiengué (YA)

Fl. Cameroun 16 (1973).

## Allophylus bullatus Radlk.

Tree, 15–18m; leaves trifoliolate; leaflets drying blackish-green above, brown below, secondary nerves 10–12 pairs, domatia conspicuous, white tufted, along midrib and secondary nerves, elliptic c. 19 × 8cm, long acuminate, cuneate, margin serrate; petiole c. 8cm; inflorescence in the leaf axils 10–21cm, branches 6–12 in the upper half to 10cm long; flowers white, 2mm. Forest; 1800m.
**Distr.:** Nigeria, Cameroon, Príncipe & São Tomé [W Cameroon Uplands (montane)].
**IUCN:** VU
**Fosimondi:** Tchiengue, 1903 4/2004.
Note: Tchiengue 1903 is tentatively placed in this taxon.

## Allophylus conraui Gilg ex Radlk.

**Syn.** *Allophylus* sp. 1 of Kupe Bakossi checklist
The plants of Kupe, Mwanenguba and the Bakossi Mountains: 398 (2004).
Shrub, to 1.5m, unarmed; main stem dark brown, raised lenticels, glabrous, leafy branches with simple, white pilose hairs; leaves trifoliolate; lateral leaflets

oblanceolate-elliptic to 13 × 5cm, acumen c. 1cm long, margin coarsely serrate, petiolule absent, central leaflet slightly broader than the laterals; petiolule to 0.5cm; petiole to 5.5(–8)cm long, white pilose; inflorescence axillary, single, 1 or 2 per leafy branch, unbranced, 2–4cm long; peduncle 0.4(–1)cm long; cymules 20–35; 10cm with 4 branches each to 2cm; pedicel c. 1.75cm long, articulated midway; flowers white; fruit red, obovoid, 8–11 × 7–8mm, appressed-puberulent, single-seeded; seed ovoid, c. 7 × 5mm. Lowland and submontane evergreen forest; 320m.
**Distr.:** Cameroon [W Cameroon Uplands].
**IUCN:** EN
**Bechati:** Tchiengue, 2787 9/2006.

## Deinbollia oreophila Cheek

Kew Bull. 64: 504 (2009).
**Syn.** *Deinbollia* sp. 1 of Kupe & Bali Ngemba
The Plants of Kupe, Mwanenguba and the Bakossi Mountains, Cameroon: 399 (2004); The Plants of Bali Ngemba Forest Reserve, Cameroon: 125 (2004).
Monopodial shrub, 0.8–3(–5)m; stems brown with bright white raised lenticels; leaves 25–63cm, (2–)3–4(–5)-jugate; leaflets pale green, nerves yellow, oblong-elliptic, c. 15–24 × 5.5–9cm, acuminate; petiole 9–16.5cm; panicle terminal, 8–20 × 5–20cm; flowers white, 3–4mm, glabrous; fruit orange, all but 1 carpel aborting, globose, 3cm diam. Understorey of submontane evergreen forest; 1400–1540m.
**Distr.:** Cameroon & SE Nigeria [W Cameroon Uplands].
**IUCN:** VU
**Fosimondi:** Tchiengue, 1950 4/2004; 2640 2/2006.

## Deinbollia sp. of Fosimondi

Subshrub, to 1m; leaflets ovate, 8–14 × 3.5–6.5cm, glabrous; fruits orange. Forest understorey; 320m.
**Distr.:** Cameroon (Fosimondi).
**Bechati:** Tchiengue, 2795 9/2006.
Note: Tchiengue 2795 is too poor to name with certainty but does not appear to fit *D. oreophila* nor any other species. When better material is available it may prove to be a new species to science, Cheek, i.2010.

# SAPOTACEAE

Y.B. Harvey (K), B. Tchiengué (YA) & X. van der Burgt (K)

Fl. Cameroun 2 (1964).

## Baillonella toxisperma Pierre

Tree, to 12m tall; c. 80cm dbh; bole nearly black, ridged, hard; slash yellow, white exudate; leaves clustered at ends of branches; stipules lanceolate; blades 20–30 × 6–10cm, rounded with acuminate apex, cuneate at base, young leaves with chestnut pubescence, subglabrous when mature although hairs persistent on midrib; inflorescence of dense flowering fascicles at the

branch tips; pedicels 2–3cm, pubescent; calyx c. 1cm long, with 8 lobes, 4 inner and 4 outer, pubescent on exterior surface; corolla with 8 lobes, each with 2 dorsal appendages longer than the lobes (5.5mm); tube 2.5mm long; lobes c. 4mm long; fruits large, spherical, c. 6.5cm diam., grey-green; 1–2-seeded in a yellowish-white pulp; seeds ellipsoid, c. 4.2 × 2.5 × 2cm, ventral scar nearly the entire length of the convex ventral face. Evergreen forest & secondary forest growth; 350m.
**Distr.:** Nigeria to W Congo (Kinshasa) [lower Guinea].
**IUCN:** VU
**Bechati:** <u>Tchiengue, 2761</u> 9/2006.
Note: the identification of Tchiengue 2761 has not been confirmed, Harvey, ii.2010.

### *Chrysophyllum* sp. of Fosimondi
Tree, to 30m; white exudate; leaves to 9cm; petiole 1cm; blade elliptic, 8cm, glabrous, apex narrowly acuminate for c. 1cm. Submontane forest; 1540m.
**Distr.:** Cameroon (Fosimondi).
**Fosimondi:** <u>Tchiengue, 2613</u> 2/2006.
Note: Tchienge 2613 comprises leaves only.

### *Synsepalum cerasiferum* (Welw.) T.D.Penn.
Pennington T., Gen. of Sapot.: 248 (1991).
**Syn.** *Afrosersalisia cerasifera* (Welw.) Aubrév.
Tree or shrub, to 40m; exudate white; leaves dark green, oblanceolate to obovate, 6–20(–35) × 5–6(–14)cm, coriaceous, obtuse, midrib and nerves raised on lower surface; petioles to 10mm; corolla greenish or yellowish-cream; tube to 3mm; lobes to 3mm, reflexed in maturity; ovary to 2mm, subconical; fruit a red ovoid to globose berry, to 3.2 × 2cm; seed to 2 × 1.5cm. Forest and farmland; 580m.
**Distr.:** Guinea (Conakry), Ivory Coast, Sierra Leone, Cameroon, Central African Republic, Gabon, Uganda, Kenya, Tanzania, Malawi, Mozambique, Zambia, Angola [tropical Africa].
**IUCN:** LC
**Bechati:** <u>van der Burgt, 884</u> 9/2006.

### *Tridesmostemon omphalocarpoides* Engl.
Tree, c. 25–30m tall; stems with numerous prominent buds in old leaf scars; outer slash thin, white, streaked red, inner yellowish white; Terminalia-style branching; leaves 9.5–19(–25) × 3–6(–8.5)cm, obovate-oblong, apex acuminate, base obtuse, glabrous; petiole 1.5–3cm, glabrous; inflorescence either solitary or paired at the base of petioles; sepals red, to 6mm long; petals oblong, joined at the base, truncate apices; staminodes 6.5–7mm long, 3-toothed; fruits held below on branches, subspherical, 10–12cm diam. at maturity, 10-seeded; seeds 4 × 1.8 × 1cm. Forest; 1300m.
**Distr.:** Cameroon to Congo (Kinshasa) [Guineo-Congolian].
**IUCN:** LC
**Fosimondi:** <u>Tchiengue, 2622</u> 2/2006.

# SCROPHULARIACEAE
S. Ghazanfar (K), B. Tchiengué (YA) & E. Fischer (University of Koblenz, Germany)

### *Lindernia nummulariifolia* (D.Don) Wettst.
Erect herb, 3–14cm; stems shortly hairy to subglabrous; leaves ovate to subcircular, 9–22 × 5–15mm, subsessile; flowers solitary in axils of upper leaves; pedicel c. 1mm; corolla blue to purple, to 6mm, lobes unequal; capsule ovoid-cylindric, to 6 × 3mm, far exceeding calyx. Roadside and streamside; 320m.
**Distr.:** Sierra Leone to Zambia & tropical Asia [palaeotropics].
**IUCN:** LC
**Bechati:** <u>Tchiengue, 2776</u> 9/2006.

### *Lindernia senegalensis* (Benth.) Skan
Annual herb; stems mostly prostrate, sometimes 30cm long, glabrous; leaves ovate to ovate-orbicular, to 2cm; petiole 2mm; pedicel 3mm; flowers axillary, solitary; corolla white, with 2 yellow palate bumps, 7mm; capsule cylindrical, 2cm, far exceeding calyx. Roadside and paths; 1486m.
**Distr.:** Senegal to Cameroon [upper & lower Guinea].
**IUCN:** LC
**Fosimondi:** <u>Tchiengue, 2202</u> 4/2005.

### *Rhabdotosperma densifolia* (Hook.f.) Hartl
Beitr. Biol. Pflanzen 53(1): 58 (1977).
**Syn.** *Celsia densifolia* Hook.f.
Robust erect herb, to 90cm; stems pithy, woody at base, tomentose; leaves tomentose beneath, lanceolate, closely serrate, 2–7 × 0.7–2cm; inflorescence a terminal raceme; pedicels 1.5cm in fruit; flowers yellow, c. 2cm diam.; fruits 6–8mm. Grassland; 1800m.
**Distr.:** Bioko, W Cameroon [W Cameroon Uplands].
**IUCN:** NT
**Fosimondi:** <u>van der Waarde, 15</u> 10/2004.

### *Veronica abyssinica* Fresen.
Prostrate, creeping herb, usually drying dark brown; stem branched from the base, pilose; leaves opposite, petiolate, ovate, serrate except towards base, 2–4 × 1–2cm; inflorescence a slender axillary peduncle; flowers blue or pinkish, paired or a few together, 8–10mm diam.; fruit bilobed, pubescent. Grassland, forest-grassland transition & roadsides; 1610m.
**Distr.:** Nigeria to Zimbabwe [afromontane].
**IUCN:** LC
**Fosimondi:** <u>Tchiengue, 2581</u> 2/2006.

# SCYTOPETALACEAE
B. Tchiengué (YA)

Fl. Cameroun 20 (1978).

### *Rhaptopetalum geophylax* Cheek & Gosline
Kew Bull. 57(3): 662 (2002).

Tree, 6–10m; stems terete, glabrous; leaves obovate-oblong (14–)21–28 × 11–15cm, acumen 1cm, base unequally rounded, glabrous; inflorescence ramiflorous, subfasciculate; pedicel 3–6mm, articulated; flowers pink, buds 9–15mm, ovary superior; fruit orange, ovoid, 27–38mm, receptacle covered in sinuous corky ridges. Forest; 1370m.
**Distr.:** Cameroon [W Cameroon Uplands].
**IUCN:** NT
**Fosimondi:** Tchiengue, 2582 2/2006.

## SIMAROUBACEAE

M. Cheek (K) & B. Tchiengué (YA)

### Brucea antidysenterica J.F.Mill.
Shrub, or tree, to 10m; leaves imparipinnate, 10–35cm long; leaflets 4–5, oblong-ovate to ovate-lanceolate, rusty tomentose beneath, 4.5–14 × 2–7cm, margins undulate; inflorescence an elongated panicle to 35cm long; flowers clustered, subsessile, green. Forest.
**Distr.:** tropical Africa [afromontane].
**IUCN:** LC
**Fosimondi:** Tchiengue, 2166 4/2005.

### Quassia sanguinea Cheek & Jongkind
Kew Bull. 63(2): 249 (2008).
**Syn.** *Hannoa ferruginea* Engl.
Tree, (1.5–)2–4(–6)m, glabrous; 3–(–4)-jugate, rachis and midrib violet, leaflets usually drying green below, c. 13 × 5cm, apex subacuminate. Forest; 1370–1500m.
**Distr.:** SE Nigeria & Cameroon [W Cameroon Uplands].
**IUCN:** VU
**Fosimondi:** Tchiengue, 2270 4/2005; 2585 2/2006.

## SOLANACEAE

B.J. Pollard, B. Tchiengué (YA) &
Y.B. Harvey (K)

### Browallia americana L.
Sp. Pl. 631 (1753).
Weedy herb, to c. 50cm; stems herbaceous, sub-woody at base, several-branched; leaves ± ovate, obtuse, cuneate, rounded or truncate, entire, shortly ciliate on nerves, c. 3–6 × 2–3.5cm; flowers borne singly along the stems, interspersed between leaves; pedicels c. 0.5–1.0cm; flowers showy, bright bluish-purple; corolla rim lobed; calyx campanulate in fruit; fruit c. 0.5 × 0.5cm. Roadside verges, path edges, farmbush; 1610m.
**Distr.:** C & S America, Caribbean, naturalised in tropical Africa, Malaya and New Guinea [pantropical].
**IUCN:** LC
**Fosimondi:** Tchiengue, 2568 2/2006.
Note: Tchiengue 2568 has white flowers.

### Cyphomandra betacea (Cav.) Sendtn.
Tree, to 6m; leaves alternate, ovate, to c. 20 × 15cm, base cordate; petiole c. 10cm long; inflorescence few-flowered, axillary, pendulous; flowers campanulate, c. 1.5 × 0.7cm wide; petals pink; fruit ellipsoid, c. 7 × 4cm, orange, edible. Forest; 1600m.
**Distr.:** native to Peru, but naturalised in many parts of the tropics [pantropical].
**IUCN:** LC
**Fosimondi:** Tchiengue, 2197 4/2005.
**Uses:** FOOD — Infructescences: fruits eaten (Tchiengue 2197). **Local name:** Tomato (Banwa) (Tchiengue 2197).

### Discopodium penninervium Hochst.
Small tree or shrub, 5–7m; leaves elliptic to oblong-elliptic, mostly glabrous, lateral nerves in 10–12 pairs, 10–25 × 3–10cm; flowers white or yellowish, fading to brown, fasciculate, axillary; corolla cylindrical; lobes reflexed or spreading, c. 8mm; berry globose, 6–8mm diam. Forest, forest edges; 1800m.
**Distr.:** Bioko, Cameroon, Congo (Kinshasa), Ethiopia & E Africa [afromontane].
**IUCN:** LC
**Fosimondi:** Tchiengue, 2254 4/2005.
**Local name:** Tapang (Banwa) (Tchiengue 2254).

### Physalis peruviana L.
Erect or straggling perennial, to 1m, densely hairy; from creeping rootstock; leaves rhomboid to deltoid, entire or with a few large teeth, 8–10 × 6–7.5cm; flowers yellow with purple centre, 15mm; fruiting calyx large, to 4 × 3cm, villous. Fallow, rocky grassland; 1800m.
**Distr.:** tropical America, naturalised in W Africa [pantropical].
**IUCN:** LC
**Fosimondi:** Tchiengue, 1969 4/2004.

### Solanum terminale Forssk.
**Syn.** *Solanum terminale* Forssk. subsp. *inconstans* (C.H.Wright) Heine
Bothalia 25: 49 (1995).
**Syn.** *Solanum terminale* Forssk. subsp. *sanaganum* (Bitter) Heine
Fl. Rwanda 3: 382 (1985).
Woody climber, to c. 10–15m; leaves elliptic, c. 12 × 5cm, acuminate, glabrous; petiole 1–2cm; inflorescence terminal or lateral, paniculate or cymose, very rarely spicate, 10–20cm; flowers c. 5 × 8mm; petals purple; staminal tube yellow. Forest; 1700–1800m.
**Distr.:** tropical Africa.
**IUCN:** LC
**Fosimondi:** Tah 291 4/2004; Tchiengue, 1924 4/2004.
Note: Tchiengue 1924 is tentatively placed in this taxon by Pollard, 2005.

### Solanum welwitschii C.H.Wright
Bothalia 25: 49 (1995).
**Syn.** *Solanum terminale* Forssk. subsp. *welwitschii* (C.H.Wright) Heine
Woody climber, ± glabrous, to c. 15m; leaves elliptic, oblong to obovate, 7–15 × 3–6cm; inflorescences

spicate and terminal; pedicels c. 3–5mm in flower, to 1.5cm in fruit; flowers purple; fruits c. 5mm, globose, reddish at maturity, smooth. Forest.
**Distr.:** Guinea (Conakry) to Bioko, Cameroon, Gabon, Congo (Brazzaville & Kinshasa) & Angola [Guineo-Congolian].
**IUCN:** LC
**Fosimondi:** Tchiengue, 2232 4/2005.

# STERCULIACEAE

## M. Cheek (K)

### *Cola anomala* K.Schum.
Tree, 15–20m; crown dense; stems bright white from waxy cuticle; leaves in whorls of 3(–4), simple, entire, elliptic, to 17 × 7.5cm, subacuminate, base rounded to obtuse, lateral nerves c. 7 pairs; petiole c. 2cm; stipules caducous; panicles 3cm in leaf axils; flowers yellow without red markings, 1.5cm; fruit follicles to 12cm, 2-seeded, with knobs and ridges. Forest; 1150–1800m.
**Distr.:** Cameroon & Nigeria [W Cameroon Uplands].
**IUCN:** NT
**Fosimondi:** Tchiengue, 2262 4/2005.
Note: Tchiengue 2262, field determined as *Cola verticillata*, although not seen, is likely to be this taxon, along with the sight records mentioned in Tchiengue habitat notes (iv.2005 & ii.2006).

### *Cola ficifolia* Mast.
Monopodial shrub or small tree, 3(–6)m; leaves simple, heteromorphic on one stem, entire, oblong, c. 30 × 11cm, truncate-caudate, acute, sub-silvery below or orbicular in outline, 3-lobed, by 3/4 incised; petiole variable, 2–20cm on one stem; stipules c. 2 × 0.4cm, scurfy; inflorescences sessile, on stem below leaves. Forest; 470m.
**Distr.:** SE Nigeria to Congo (Kinshasa) [lower Guinea & Congolian].
**IUCN:** LC
**Bechati:** van der Burgt, 858 9/2006.

### *Cola flaviflora* Engl. & K.Krause
Small tree, 2–8m; leaves orbicular in outline, to 35cm, digitately 3-lobed, green below; petiole to 20cm; inflorescence axillary, sessile; flowers urceolate, yellow-green, <1cm; fruit follicles elongate, each 6 × 2.5 × 2cm, bright red. Forest; 615m.
**Distr.:** SE Nigeria & W Cameroon [lower Guinea].
**IUCN:** NT
**Bechati:** van der Burgt, 893 9/2006.

### *Cola lepidota* K.Schum.
Tree, 5–20m, glabrescent; leaves trifoliolate, leaflets oblanceolate-obovate, 30 × 12cm, acumen 2cm, acute, densely silvery lepidote below; petiolule 1cm; petiole to 30cm; stipules caducous; cauliflorous, inflorescences to 10cm diam., with numerous 1–5-flowered peduncles to 4cm; perianth red, cup-shaped, 1cm; fruit oblong, red, 10 × 3cm. Forest; 540m.

**Distr.:** SE Nigeria, Cameroon & Gabon [lower Guinea].
**IUCN:** LC
**Bechati:** Sight record, 15 9/2006.
Note: known only from habitat record (Tchiengue 2833), no collection made, Harvey, ii.2010.

### *Cola verticillata* (Thonn.) Stapf ex A.Chev.
Tree, 6–25m tall. Forest; 1120-1920m.
**Distr.:** Ghana to Congo (Kinshasa) and Angola (Cabinda) [Guineo-Congolian].
**IUCN:** LC
**Fosimondi:** Sight record, 16 4/2005; 2/2006.
Note: the habitat record sightings have not been confirmed, Harvey, ii.2010.

### *Leptonychia kamerunensis* Engl. & K.Krause
Kew Bull. (in prep.); Sterculiaceae Africanae V. Bot. Jahrb. Syst. 45: 322 (1911).
**Syn.** *Leptonychia* sp. 1 of Bali Ngemba checklist
The Plants of Bali Ngemba Forest Reserve, Cameroon: 129 (2004).
**Syn.** *Leptonychia* sp. 1 of Kupe-Bakossi checklist
The Plants of Kupe, Mwanenguba and the Bakossi Mountains, Cameroon: 411 (2004).
Tree, to 17m, 30cm dbh; leaves elliptic or rhombic-oblong, to 23 × 8cm, acumen slender, to 2cm, obtuse, 3-nerved; petiole 1.5cm; panicles subfasciculate, axillary, 1–5-flowered, 3cm; flowers 1cm, dull white; anthers in 5 bundles; fruit globose, 4cm, minutely rugose, densely brown puberulent; seed 3cm. With *Santiria trimeria* (Burseraceae), *Cola verticillata* (Sterculiaceae), *Macaranga occidentalis* (Euphorbiaceae); 1370m.
**Distr.:** Cameroon, SW & NW Region [Cameroon Endemic].
**IUCN:** EN
**Fosimondi:** Tchiengue, 2208 4/2005.
**Uses:** FOOD — Infructescences: the fruits can be stewed like those of okra (Tchiengue 2208). **Local name:** Keliteh (Banwa) (Tchiengue 2208).
Note: Tchiengue 2208 not seen at Kew so identification uncertain, but no other species of the genus occurs at this altitude in the Cameroon Highlands, Cheek, ii.2010.

### *Sterculia tragacantha* Lindl.
Tree, 10–25m; leaves simple oblong, oblong-obovate or elliptic, c. 19 × 11cm, rounded to subacuminate, base obtuse, stellate-velvety below; petiole c. 5cm; stipules caducous; panicles axillary, slender, 15cm; flowers pink, 0.7cm; perianth lobes adhering at apex; follicles five c. 8.5 × 2.5cm, golden-brown hairy. Secondary forest; 320m.
**Distr.:** Mali to Mozambique [tropical Africa].
**IUCN:** LC
**Bechati:** Sight record, 17 9/2006.
Note: known only from habitat record (Tchiengue 2789, ix.2006), Harvey, ii.2010.

# THYMELAEACEAE

Y.B. Harvey (K)

Fl. Cameroun 5 (1966).

### *Dicranolepis vestita* Engl.
Shrub or tree, (2–)3–8m; leaves obliquely oblong-elliptic, c. 8 × 2.5cm, acumen 1.5cm; flowers erect, sessile, 1–several in old leaf axils; flowers 2.5cm wide, white, fragrant; calyx tube c. 2cm, densely white appressed-hairy, dilated at base; petals 10, entire; anthers sessile; stigma capitate; fruit pendulous, 2.5 × 1.5cm including a 1cm robust rostrum, densely white hairy. Forest; 1800m.
**Distr.:** SE Nigeria, Bioko & Cameroon [lower Guinea].
**IUCN:** NT
**Fosimondi:** Tchiengue, 1910 4/2004.

### *Gnidia glauca* (Fres.) Gilg
Fl. Gabon 11: 95 (1966); Fl. Cameroun 5: 69 (1966); Fl. Afr. Cent. Thymelaeaceae: 61 (1975).
**Syn.** *Lasiosiphon glaucus* Fresen.
Tree, to 15m; trunk much-branched; leaves oblanceolate, very acute, glabrous, 5–8cm long; flowerheads numerous, subsessile, c. 5cm diam.; petals spathulate, surrounded by large ovate glabrescent bracts. Open woodland, forest-grassland transition.
**Distr.:** Cameroon, Ethiopia, Zambia and E Africa [afromontane].
**IUCN:** LC
**Fosimondi:** Sight record, 18 4/2005; 2/2006.
Note: known only from sight record (Tchiengue 2165, iv.2005 & photograph, ii.2006), Harvey, ii.2010.

### *Peddiea africana* Harv.
**Syn.** *Peddiea fischeri* Engl.
**Syn.** *Peddiea parviflora* Hook.f.
Shrub, or tree, 3–7m; leaves alternate, elliptic, c.15 × 5cm, base and apex acute; petiole 0.3cm; inflorescence terminal; peduncle 0.8cm, umbellate, c. 10-flowered; pedicel 0.5cm; perianth tubular, pale green, 1.8cm; lobes 4, 0.2cm; fruit ovoid, 1 × 0.7cm, bilobed, apex white hairy. Forest; 1450m.
**Distr.:** Guinea (Conakry) to Zambia [afromontane].
**IUCN:** LC
**Fosimondi:** Tchiengue, 2233 4/2005.
Note: Tchiengue 2233 has not been verified (field determination: *Peddiea fischeri*), Harvey, ii.2010.

# TILIACEAE

M. Cheek (K) & X. van der Burgt (K)

### *Desplatsia subericarpa* Bocq.
Shrub 3–6m, slender, dark brown pubescent; leaves c. 17 × 6cm, obscurely serrate, glabrous below; stipule 5mm, digitately divided; flowers pink, 1cm; fruit c. 5 × 5cm. Forest; 350m.
**Distr.:** Sierra Leone to Congo (Kinshasa) [Guineo-Congolian].

**IUCN:** LC
**Bechati:** Tchiengue, 2766 9/2006.

### *Microcos barombiensis* (K.Schum.) Cheek
The Plants of Kupe, Mwanenguba and the Bakossi Mountains: 414 (2004).
**Syn.** *Grewia barombiensis* K.Schum.
Climber, stems brown puberulent; leaves ovate-elliptic, c. 17 × 9cm, acuminate, rounded or cordate, entire, subglossy above, glabrous below apart from nerves; stipules c. 6mm, bifurcate; inflorescence terminal, 10 × 7cm; main bracts digitately divided to base; flowers in umbels of 2–3, subtended by 5–6 epicalycular bracteoles; flowers white, 13mm diam.; fruit spindle-shaped, 3 × 1.5cm glossy red, with white thinly scattered hairs. Forest; 680m.
**Distr.:** Ivory Coast to Angola [upper & lower Guinea].
**IUCN:** NT
**Bechati:** van der Burgt, 895 fl., 9/2006.
Note: *Microcos* is not always maintained as distinct from *Grewia* in Africa despite Burret's excellent revisionary work of 1926 (Notizbl. Bot. Gart. Berlin 9).

### *Microcos coriacea* (Mast.) Burret
Notizbl. Bot. Gart. Berlin 9: 759 (1926).
**Syn.** *Grewia coriacea* Mast.
Evergreen, tree c. 20m; leaves ovate-elliptic, 30 × 9–14cm, acuminate, obtuse to rounded, entire, glabrous below; petiole 1cm; inflorescence a terminal panicle; fruit obovoid, glossy, 2.5 × 2cm, red, fleshy, 1-seeded. Evergreen forest.
**Distr.:** S Nigeria to Angola [lower Guinea & Congolian].
**IUCN:** LC
**Fosimondi:** Sight record, 19 2/2006.
Note: tentatively placed in this taxon from label habitat notes (*Microcos* cf. *coriacea*) (Tchiengue 2625–2633 & 2636–2637), no collections made, Harvey, ii.2010.

# ULMACEAE

B. Tchiengué (YA)

Fl. Cameroun 8 (1968).

### *Trema orientalis* (L.) Blume
**Syn.** *Trema guineensis* (Schum. & Thonn.) Ficalho
Tree, to 8m; young stems densely pubescent; leaves variable, distichous, ovate-lanceolate, 6.5–13.5 × 2.8–5.3cm, apex acuminate, base truncate, margin serrulate, lateral nerves 4(–6) pairs, alternate above basal pair, upper surface scabrid, lower surface scabrid or sparsely pubescent to densely pubescent; cymes axillary, c. 10–20-flowered; peduncle 0–0.5cm; flowers white, c. 2mm; sepals broadly elliptic, obtuse, puberulent; fruit globose, 2–3mm diam., green, styles and sepals persistent. Forest & farmbush.
**Distr.:** widespread in tropical Africa & Asia [palaeotropics].
**IUCN:** LC
**Fosimondi:** Sight record, 20 4/2004.

## UMBELLIFERAE

### B. Tchiengué (YA)

Fl. Cameroun 10 (1970).

**Sanicula elata** Buch.-Ham.
Upright herb, 0.5(–1)m; short stolon at base; upright stems c. 2mm diam., glabrous, ridged; basal rosette of 2–4 leaves on petioles, to 15cm, 3(–5)-lobed almost to base, with irregular sublobing and dentate-mucronate margins, c. 6 × 8cm, glabrous; cauline leaves smaller on petioles, to 5cm; inflorescence cymose, bifurcating 3–4 times with a central cyme on peduncle, 1–1.5cm long; lateral primary peduncles 5–6cm, secondary peduncles c. 1–2cm; each floret of 2–3 sessile flowers; florets and peduncles subtended by lanceolate bracts, 6–12mm; flowers 1–2mm; corolla white; fruits ellipsoid, 3 × 2mm, covered in hooked bristles, green. Forest including stream edges; 1350–1540m.
**Distr.:** tropical & South Africa, Madagascar, Comores & temperate Asia [montane].
**IUCN:** LC
**Fosimondi:** Tchiengue, 2242 4/2005; 2594 2/2006.

## URTICACEAE

### I. Friis (C)

Fl. Cameroun 8 (1968).

**Boehmeria macrophylla** Hornem.
F.T.E.A. Urticaceae: 44 (1989); Fl. Zamb. 9(6): 108 (1991).
**Syn. Boehmeria platyphylla** D.Don
Shrub, to 2(–3)m; branches glabrous, except when young; leaves opposite, anisophyllous, ovate, 10–13.5 × 5.5–9cm, acuminate, base acute to rounded, margin serrate, basal lateral nerves prominent, upper surface sparsely pubescent, cystoliths punctiform, lower surface glabrescent; petiole to 6cm; spikes axillary, 7–50cm, whip-like, with glomerules of flowers spaced 1–10mm apart; male glomerules 1–2mm; female 2–3mm. Forest & forest edge; 320–1580m.
**Distr.:** tropical Africa & Madagascar, tropical Asia to SW China [palaeotropics].
**IUCN:** LC
**Bechati:** Tchiengue, 2784 9/2006; **Fosimondi:** Tchiengue, 2200 4/2005.

**Elatostema welwitschii** Engl.
Herb, to 30cm, occasionally branched; leaves drying fresh green, marginal teeth fine, 22–29(–40) distally, 16–24 proximally, sometimes biserrate, tips ciliate, sparsely pubescent in midrib below, distal base rounded; stipules to 7mm; inflorescence 0.4–0.8cm wide, dense, pale green. Rocks by forest streams; 1800m.
**Distr.:** Nigeria to Tanzania & Malawi [afromontane].
**IUCN:** LC
**Fosimondi:** Tchiengue, 1960 4/2004; 1961 4/2004.

**Pilea rivularis** Wedd.
Fl. Cameroun 8: 163 (1968); F.T.E.A. Urticaceae: 29 (1989); Fl. Zamb. 9(6): 98 (1991).
**Syn. Pilea ceratomera** Wedd.
Erect herb, to 60cm; stems with linear cystoliths; stipules prominent, oblong, 7mm; leaves to 7.5 × 5cm, base rounded, margin serrate-crenate, cystoliths linear; inflorescence a dense axillary cluster to 1.5cm diam. Forest including rocky stream margins; 1400m.
**Distr.:** tropical & subtropical Africa [afromontane].
**IUCN:** LC
**Fosimondi:** Tchiengue, 2639 2/2006.

**Urera gravenreuthii** Engl.
Liana to 8m; stems densely covered in gold/bronze coloured hairs; leaves elliptic, 12–16 × 7–9cm, 5–7 nerves on each side of the midrib, dentate, adaxial surface dark green with scabrid hairs, abaxial surface paler green, pubescent (particularly along the veins); petiole 1.5–8cm; panicle 5–7cm; flower buds 1mm diam., red-brown; fruit clusters on leafless parts of the stem. Forest; 320–1386m.
**Distr.:** Cameroon [lower Guinea (montane)].
**IUCN:** NT
**Bechati:** Tchiengue, 2804 9/2006; **Fosimondi:** Tchiengue, 2196 4/2005.
**Local name:** Ndab (Banwa) (Tchiengue 2196).

## VIOLACEAE

### Y.B. Harvey (K) & B. Tchiengué (YA)

**Rinorea dentata** (P.Beauv.) Kuntze
Shrub or small tree, to 5m; stems puberulent; leaves papery, elliptic, 15.5–19.5 × 6–8cm, acumen 1.7cm, base acute, margin denticulate, lateral nerves 11–13 pairs, midrib puberulous below, laminae glabrous except when young, eglandular; petiole 0.6–1.2cm; panicles terminal, c. 3cm, puberulent, c. 8–10-flowered; peduncle 3cm; sepals triangular, 1.5mm; petals lanceolate, 4mm, yellow. Forest; 320m.
**Distr.:** Liberia to Uganda [Guineo-Congolian].
**IUCN:** LC
**Bechati:** Tchiengue, 2794 9/2006; van der Waarde, 3 10/2004.
Note: van der Waarde No. 3 has tentatively been placed in this taxon. Many of the diagnostic features are not visible in the photograph (specimen not seen), Harvey, ii.2010.

**Rinorea preussii** Engl.
Tree, 2–12m; stems glabrous, pale; leaves (oblong-) elliptic, 14–21 × 6–11cm, acumen 1.5cm, base acute to obtuse, margin serrulate, lateral nerves prominent, 11–12 pairs, glabrous, eglandular; petioles 2.5–5cm; panicles terminal, elongate, to 20cm, somewhat lax, finely puberulent, many-flowered; sepals rounded, 2.5mm; petals ovate, 4.5mm, white to pale yellow; fruit glabrous. Forest & forest edge; 1350–1500m.

**Distr.:** Liberia & SW Cameroon [lower Guinea & Congolian].
**IUCN:** NT
**Fosimondi:** <u>Tah 303</u> 4/2004; <u>317</u> 4/2004; <u>Tchiengue, 2590</u> 2/2006.
Note: Tah 303 & 317 have tentatively been placed in this taxon.

# VITACEAE

L. Pearce (K) & Y.B. Harvey (K)

Fl. Cameroun 13 (1972).

## *Cyphostemma adenopodum* (Sprague) Desc.
Fl. Cameroun 13: 46 (1972); F.T.E.A. Vitaceae: 131 (1993).
**Syn.** *Cissus adenopoda* Sprague
Herbaceous climber, (reddish) pubescent with long jointed and glandular hairs; stems cylindrical; tomentose with a few glandular hairs; tendrils bifid; leaves palmately 3-foliolate; leaflets thinly membranous, 6–11 × 3–6cm; petiolules to 1cm; central leaflet elliptic to obovate, apex shortly acuminate, base cuneate or obtuse; lateral leaflets strongly asymmetrical, broadly rounded on one side at base; margins coarsley crenate-dentate; petioles 3–6cm; inflorescence a loose compound cyme, to 22cm; flowers 0.3cm; pedicels bearing approx. 1–4 glandular hairs; unexpanded corolla subcylindrical, constricted in the middle; fruits red, up to 1cm diam., glabrous; seeds globular, 3.5–5mm. Forest; 320m.
**Distr.:** Ivory Coast to Bioko, Cameroon, Gabon, Congo (Kinshasa), Angola (Cabinda) & Uganda [Guineo-Congolian].
**IUCN:** LC
**Bechati:** <u>van der Burgt, 861</u> 9/2006.

## *Cyphostemma mannii* (Baker) Desc.
Fl. Cameroun 13: 68 (1972).
**Syn.** *Cissus mannii* (Baker) Planch.
Herbaceous climber; stems cylindrical, tomentose; tendrils bifid; leaves palmately 5-foliolate, to 10.5 × 11cm; leaflets obovate-elliptic, central 5–7.0 × 2.5–3.5cm, apex shortly acuminate, base cuneate, margin crenate-serrate, tomentose along veins; petiole 2.5–3.5cm, tomentose; compound cyme c. 12 × 8cm, subcorymbiform, densely tomentose; flowers 0.4cm; calyx cupular; corolla pinched in centre, apex truncate; fruit globose, 6mm diam., glabrous; seeds ellipsoid, 5mm, ridged. Forest & thicket.
**Distr.:** Bioko & Cameroon [W Cameroon Uplands].
**IUCN:** LC
**Fosimondi:** <u>Tchiengue, 2167</u> 4/2005.
Note: the dimensions of Tchiengue 2167 are smaller than is usual for this species. It is likely however that elsewhere on the plant they were more representative. Specimen flowering. Description of fruit from Flore du Cameroun 13: 69–70 (1972), Pearce, i.2010.

## *Cyphostemma rubrosetosum* (Gilg & Brandt) Desc.
Fl. Cameroun 13: 56 (1972).
**Syn.** *Cissus rubrosetosa* Gilg & Brandt
Herbaceous climber; stems cylindrical; stem and petioles bearing jointed, red hairs to 3mm; tendrils bifid; leaves papery, palmately 5-foliolate, 11–13.5 × 9.5–11cm; leaflets obovate to elliptic, central 6–8 × 2–2.5cm, apex acuminate, base cuneate, margin crenate-dentate, pubescent, especially along veins; petiole 4.5–7cm; compound cyme, 9 × 9cm; peduncles pubescent; flower 0.4cm; calyx cupular; corolla pinched at the centre, apex truncate; pedicels bearing glandular hairs; fruit subglobose, 0.5cm diam., glabrous; seeds globose, c. 4mm, striate. Farmbush & open forest; 1800m.
**Distr.:** Guinea (Conakry) to CAR [upper & lower Guinea].
**IUCN:** LC
**Fosimondi:** <u>Tchiengue, 1919</u> 4/2004.
Note: the dimensions of Tchiengue 1919 are smaller than is usual for this species. It is likely however that elsewhere on the plant they were more representative. Please note however that it also lacks glandular hairs on the vegetative parts. Specimen flowering. Description of fruit from Flore du Cameroun 13: 69–70 (1972), Pearce, i.2010.

# MONOCOTYLEDONAE

## AMARYLLIDACEAE

M. Cheek (K)

Fl. Cameroun 30 (1987).

***Crinum jagus*** (Thomps.) Dandy
Bulbous herb, 0.6–1m; leaves erect, often petiolate, c. 50 × 7cm, c. 30-nerved, margin often undulate; inflorescence above leaves, 3-flowered, with 2 involucral bracts; flowers irregular, infundibuliform, sweetly scented, 7 × 3.5cm, white with green markings; stamens black. Forest, savanna, plantations, usually riverine; 800m.
**Distr.:** Guinea (Conakry) to Angola, Sudan, Uganda [Guineo-Congolian].
**IUCN:** LC
**Bechati:** van der Waarde, 6 10/2004.

***Scadoxus cinnabarinus*** (Decne.) Friis & Nordal
Fl. Gabon 28: 25 (1986); Fl. Cameroun 30: 5 (1987).
**Syn.** *Haemanthus cinnabarinus* Decne.
Herb, 0.4–0.5m; lacking bulb, acaulous; leaves numerous; blades elliptic, c. 20 × 8cm; petiole 14cm, winged; inflorescence terminal; peduncle 20–30cm; flowers pink, in umbel 10–15cm diam., each flower 4cm wide; fruit globose, orange. Forest; 1800m.
**Distr.:** Sierra Leone to Bioko, Cameroon, Angola & Uganda [Guineo-Congolian].
**IUCN:** LC
**Fosimondi:** Tchiengue, 1920 4/2004.

## ANTHERICACEAE

B. Tchiengué (YA)

***Chlorophytum comosum*** (Thunb.) Jacq. var. ***bipindense*** (Baker) A.D.Poulsen & Nordal
Bot. J. Linn. Soc. 148(1): 15 (2005).
Herb, to 30–40cm; lamina dark green above, lighter below, oblanceolate, c. 16–30 × 3.5–4cm, acute-mucronate, base tapering into a variably defined petiole; inflorescence longer than leaves, to 40cm, unbranched, to 12 flowering nodes per rachis; pedicel to 6mm; flowers white; capsules longer than wide, to 7mm long, truncate. Farmland; 350m.
**Distr.:** Liberia, Sierra Leone, Ivory Coast, Ghana, Nigeria, Cameroon, Gabon, Congo (Kinshasa) [Guineo-Congolian].
**IUCN:** LC
**Bechati:** van der Burgt, 856 9/2006.

***Chlorophytum comosum*** (Thunb.) Jacq. var. ***sparsiflorum*** (Baker) A.D.Poulsen & Nordal
Bot. J. Linn. Soc. 148(1): 15 (2005).
Herb, 25–60cm, drying light green, sometimes viviparous; leaves oblanceolate or oblanceolate-ligulate, c. 25 × 6cm, acute-mucronate, base tapering into a variably defined petiole; inflorescence about as long as leaves, or longer. Forest; 1450m.
**Distr.:** Sierra Leone to Kenya [afromontane].
**IUCN:** LC
**Fosimondi:** Tchiengue, 2193 4/2005.

## ARACEAE

A. Haigh (K)

Fl. Cameroun 31 (1988).

***Amorphophallus* sp. of Fosimondi**
Herb, to 0.6m; leaf like a tattered umbrella. leaflets acute, apical leaflet fishtail shaped; petiole to 0.65m, smooth; peduncle to 23cm; spathe 23 × 12mm; fruits green turning orange, 13 × 6mm. Cocoyam farm in submontane forest; 1486m.
**Distr.:** Cameroon (Fosimondi).
**Fosimondi:** Tchiengue, 2190 4/2005.

***Anchomanes difformis*** (Blume) Engl.
**Syn.** *Anchomanes difformis* (Blume) Engl. var. *pallidus* (Hook.) Hepper
**Syn.** *Anchomanes welwitschii* Rendle
Herb; leaf like a tattered umbrella, leaflets fishtail-shaped; petiole spiny (prickles); spathe green tinged purple; styles straight but very short, smooth, green. Forest & forest margins; 1300m.
**Distr.:** Guineo-Congolian.
**Fosimondi:** Tah 324 4/2004.
Note: *Anchomanes hookeri* was until recently treated as a synonym of *A. difformis* but the two are separable on differences in the styles. However, the distributions of these taxa have not been fully defined as yet; no conservation assessments can be made until this is clarified. Tah 324 is a unicate, and has not been verified at Kew, Harvey, ii.2010.

***Anubias barteri*** Schott var. ***barteri***
Meded. Land. Wag. 79(14): 12 (1979).
Terrestrial herb; leaves ovate-lanceolate, 7–20 × 4–10cm, base truncate to cordate, many prominent lateral nerves; fruits green, enclosed within a green, persistent spathe. Rivers & streams in forest; 1350m.
**Distr.:** SE Nigeria, Bioko, Cameroon [lower Guinea].
**IUCN:** LC
**Fosimondi:** Tchiengue, 2215 4/2005.
**Local name:** Koo Njimenguem (Banwa) (Tchiengue 2215).

***Colocasia esculenta*** (L.) Schott
Erect herb, with subglobose tubers to 10cm diam., not staining yellow when cut; leaves peltate, lamina cordiform, 20–40 × 15–35cm, basal lobes obtuse to rounded, 3–7cm; peduncle c. 15cm; spathe 1.8cm, acuminate; spadix much shorter. Cultivated; 1340–1600m.
**Distr.:** originating in Polynesia, widely cultivated.
**IUCN:** LC

**Fosimondi:** <u>Sight record, 1</u> 4/2005, 2/2006.
Note: known only from sight records, Harvey, ii.2010.

*Culcasia scandens* P.Beauv.
**Syn.** *Culcasia lancifolia* N.E.Br.
**Syn.** *Culcasia saxatilis* A.Chev.
Herbaceous climber, or epiphyte; stems verrucate, numerous internodal roots; leaves asymmetrically ovate, to 15cm long, translucent lines very short, lamina not decurrent into the petiole; petiole with a spreading sheath, terminating c. 5mm below the blade; spathe to 3.5cm; fruiting spadix terminating in an appendage. Forest; 1350m.
**Distr.:** Senegal to Congo (Kinshasa) [Guineo-Congolian].
**IUCN:** LC
**Fosimondi:** <u>Tchiengue, 2216</u> 4/2005.

*Culcasia striolata* Engl.
Terrestrial herb with stilt roots; lamina oblong, with numerous prominent short translucent lines. Forest; 250m.
**Distr.:** Guinea (Conakry) to Gabon [upper & lower Guinea].
**IUCN:** LC
**Bechati:** <u>Tchiengue, 2774</u> 9/2006.

*Nephthytis poissonii* (Engl.) N.E.Br.
**Syn.** *Nephthytis gravenreuthii* (Engl.) Engl.
Fl. Cameroun 31: 50 (1988).
**Syn.** *Nephthytis constricta* N.E.Br.
**Syn.** *Nephthytis poissonii* (Engl.) N.E.Br. var. *constricta* (N.E.Br.) Ntepe-Nyame
Fl. Cameroun 31: 56 (1988).
Terrestrial creeping herb; leaves triangular, posterior lobes considerably more developed than the anterior lobes; fruits orange-red subtended by a spreading, persistent green spathe. Forest; 1800m.
**Distr.:** Sierra Leone to Gabon [upper & lower Guinea].
**IUCN:** LC
**Fosimondi:** <u>Tchiengue, 1906</u> 4/2004.

## ASPARAGACEAE

J. van der Waarde

*Asparagus warneckei* (Engl.) Hutch.
Woody climber, to 8m; stems smooth, glabrous; spines short and recurved, only on the main shoots; cladodes flattened, ± falcate, c. 4 × 0.2cm; inflorescence racemose, c. 5–7cm, often clustered; pedicels jointed nearly at the base, about 0.5cm in flower; flowers white or cream-coloured with strong sickly odour. Thicket, forest & scrub; 1800m.
**Distr.:** Guinea (Conakry), Ghana, Togo, W Cameroon [upper & lower Guinea].
**IUCN:** NT
**Fosimondi:** <u>van der Waarde, 22</u> 10/2004.
Note: uncommon over much of its range.

## COLCHICACEAE

B. Tchiengué (YA)

*Gloriosa superba* L.
**Syn.** *Gloriosa simplex* L.
Herb, ± climbing; leaves sessile, in whorls of 3 or opposite or alternate, c. 8–10 × 4cm; often apically tendriliform; flowers yellow, turning red; perianth segments narrowly linear, 6–9 × c. 1cm with very crispy-waved margins; style longer than the perianth; filaments shorter; anthers c. 1–2cm long; fruits c. 9 × 2cm. Forest & farmbush; 283m.
**Distr.:** widely distributed in tropical & S Africa; tropical Asia [palaeotropics].
**IUCN:** LC
**Bechati:** <u>Tchiengue, 2772</u> 9/2006.

## COMMELINACEAE

R. Faden (US), Y.B. Harvey (K), M. Cheek (K) & B. Tchiengué (YA)

*Aneilema dispermum* Brenan
Weak erect herb, to c. 1m; leaves elliptic, to 14 × 4cm, acuminate, sessile or shortly petiolate, margin ciliate; inflorescence terminal, dense, c. 4.5 × 3cm; flowers 4–5mm, white; capsules broader than long, 2-seeded. Montane forest edge, forest-grassland transition; 1486m.
**Distr.:** Bioko, SW Cameroon, Malawi & Tanzania [afromontane].
**IUCN:** NT
**Fosimondi:** <u>Tchiengue, 2189</u> 4/2005.

*Aneilema silvaticum* Brenan
Weak, erect herb, 15–30cm tall; leaves ovate-lanceolate c. 9 × 3cm, acuminate, petiolate, margin ciliate; inflorescence terminal, dense, < 2 × 2cm; flowers <10mm, white; fruit pointed. Lowland forest; 550m.
**Distr.:** Nigeria, Cameroon & Congo (Kinshasa) [lower Guinea & Congolian].
**IUCN:** VU
**Bechati:** <u>Tchiengue, 2832</u> 9/2006.

*Coleotrype laurentii* K.Schum.
Scandent or decumbent herb, to 1m; leaves elliptic, c. 15 × 4cm, long-acuminate, base attenuate; inflorescences axillary, perforating the leaf-sheath, sessile, c. 3 × 4cm, subtended by leafy bracts; flowers white, tubular. Lowland forest; 550m.
**Distr.:** Ivory Coast to Uganda [Guineo-Congolian].
**IUCN:** NT
**Bechati:** <u>van der Burgt, 891</u> fl., fr., 9/2006.

*Commelina capitata* Benth.
Clustering, robust, erect herb, 0.3–1m tall; leaves oblong-elliptic, c. 12 × 4.5cm, long-acuminate, base obtuse, strongly oblique, sheaths with reddish brown bristles at apex; spathes 3–7, clustered, c. 1.5–2 × 1cm, margin

brown hirsute, peduncle 0.5–1cm; flowers pale yellow, open 8–11 am. Lowland to montane forest; 1300m.
**Distr.:** Senegal to Bioko, Cameroon, Uganda, Tanzania & Angola [Guineo-Congolian].
**IUCN:** LC
**Bechati:** van der Waarde, 4 10/2004; **Fosimondi:** Tchiengue, 2629 2/2006.

*Cyanotis barbata* D.Don
Erect herb, 10–30cm; with underground rootstock; leaves linear-lanceolate, to 12 × 1cm, white pubescent, sessile; spathes 4–5, c. 1 × 0.5cm, pedunculate; flowers blue, actinomorphic; filaments bearded. Montane grassland; 1800m.
**Distr.:** tropical Africa & the Himalayas [palaeotropics (montane)].
**IUCN:** LC
**Fosimondi:** van der Waarde, 18 10/2004.

*Floscopa africana* (P.Beauv.) C.B.Clarke subsp. *petrophila* J.K.Morton
Straggling herb, to 20cm; leaves to three times as long as broad, 2.5–6cm long. Lowland farmbush & forest edge, sometimes on rocks; 1300m.
**Distr.:** Liberia to Uganda [Guineo-Congolian].
**IUCN:** LC
**Fosimondi:** Tchiengue, 2630 2/2006.

*Palisota mannii* C.B.Clarke
Herb, 20–80cm; lacking aerial stem; leaves forming a basal rosette, lanceolate or lanceolate-obovate, 25–40 × 5–9cm, apex acuminate, base cuneate, margin hairy, lower surface white; inflorescence cylindrical, c. 12–18 × 3.5cm; peduncle 10–50cm long; pedicels longer than flowers; flowers white; fruits red. Forest & forest-grassland transition; 1370–1700m.
**Distr.:** S Nigeria to S Sudan, Uganda & W Tanzania [lower Guinea & Congolian].
**IUCN:** LC
**Fosimondi:** Tchiengue, 1945 4/2004; 2185 4/2005; 2210 4/2005.
**Local name:** Azeze (Banwa) (Tchiengue 2185); Azezeh (Banwa) (Tchiengue 2210).

*Stanfieldiella imperforata* (C.B.Clarke) Brenan var. *imperforata*
Straggling herb; leaves glabrous; inflorescences lax, diffuse, 2–6cm long; capsules exceeding sepals; seeds smooth. Lowland to submontane forest; 320m.
**Distr.:** Sierra Leone to Ethiopia, Uganda & Tanzania [Guineo-Congolian].
**IUCN:** LC
**Bechati:** Tchiengue, 2806 9/2006.

*Stanfieldiella oligantha* (Mildbr.) Brenan var. *oligantha*
Straggling herb; leaves glabrous; inflorescence dense, branched, c. 1.5cm long; capsules exceeding sepals; seeds verrucose. Lowland & submontane forest; 1290m.

**Distr.:** Liberia to Cameroon [upper & lower Guinea].
**IUCN:** LC
**Fosimondi:** Tchiengue, 2631 2/2006.

## COSTACEAE

B. Tchiengué (YA)

Fl. Cameroun 4 (1965).

*Costus letestui* Pellegr.
Fl. Cameroun 4: 86 (1965).
Epiphytic herb; stems pendulous to 2(–6)m; leaves elliptic, c. 15 × 4cm; ligule cylindrical, truncate, 3cm or more; inflorescence axillary; flowers 5; bracts shorter than calyces; flowers pale pink; labellum pink with yellow centre. Forest; 1370m.
**Distr.:** Cameroon & Gabon [lower Guinea].
**IUCN:** NT
**Fosimondi:** Tchiengue, 2592 2/2006.

## CYPERACEAE

D. Simpson (K) & B. Tchiengué (YA)

*Carex echinochloë* Kunze
Tufted leafy perennial, to ± 1m; rhizome creeping; stems 50–100cm, distinctly triangular below, rounded-triangular above; largest leaf-blades 40–120 × 0.6–1.4cm, flat or ± plicate; inflorescence a slender, very much branched panicle, 20–50cm long, widening with age; spikelets 5–10 × 5mm, male above and female below; glumes 4–5mm. Grassland & forest edges; 1460m.
**Distr.:** W Cameroon, E & NE Africa [afromontane].
**IUCN:** LC
**Fosimondi:** Tchiengue, 2275 4/2005.

*Cyperus fertilis* Boeck.
Tufted annual herb; stem 2–10cm; 5–10 basal leaves distinctly wider above than below; inflorescence with 2–8 major peduncles of unequal length, up to 40cm, each triangular, flattened, carrying 1–3 spikelets; the peduncles sometimes recurved and proliferating (viviparous); involucral bracts leafy, elliptical, 5–10 per culm, to 20cm; spikelets light brown, flattened, 5–10 × 3–5mm. Roadsides, villages & damp places in forest; 400m.
**Distr.:** Liberia to Angola & Uganda [Guineo-Congolian].
**IUCN:** LC
**Bechati:** van der Burgt, 877 9/2006.

*Cyperus tomaiophyllus* K.Schum.
Haines R. & Lye K., Sedges & Rushes of E Africa: 207 (1983).
**Syn.** *Mariscus tomaiophyllus* (K.Schum.) C.B.Clarke
Very robust perennial; rhizome woody, branching, to 2cm thick; inflorescence umbell-like, 6–15cm wide; major spikes 6–12; peduncles 0.5–4 (rarely to 15)cm.

Swamp, grassland & forest.
**Distr.:** tropical Africa [afromontane].
**IUCN:** LC
**Fosimondi:** Tchiengue, 2234 4/2005.

### *Hypolytrum heteromorphum* Nelmes
Stout rhizomatous perennial herb; culms slender, 25–40cm × 1–2mm; leaves densely set near base, 30–60 × 1.0–1.5cm, flat or ± plicate, margin slightly toothed; inflorescence a series of panicles, broader than tall; panicles of a few closely-set major branches arising from the axils of the main bracts; bracts greyish-brown, to 3.5cm; lowest branches to 2.5cm, with 2–6 sessile spikes at the tip; spikes 1.0–1.5 × 0.1–0.2mm, brown. Wet forest; 590m.
**Distr.:** tropical Africa [afromontane].
**IUCN:** LC
**Bechati:** Tchiengue, 2820 9/2006.

## DRACAENACEAE

### G. Mwachala (EA)

### *Dracaena arborea* Link
Tree, 10–20m, trunk 30cm, with aerial roots, several-branched; leaves in dense heads, sword-shaped, 50–120 × 4.5–6cm, widest above the middle, apex acute, mucro to 3mm, base clasping stem for $^3/_4$ circumference; inflorescence pendulous, to 1.5m; perianth cream-white, c. 1.5cm; fruit to 2cm, orange-red. Submontane forest & planted.
**Distr.:** Sierra Leone to Angola [Guineo-Congolian].
**IUCN:** LC
**Fosimondi:** Photographic record, 2 4/2005.
Note: known only from a photographic record, Harvey, ii.2010.

### *Dracaena camerooniana* Baker
Shrub, 0.3–8m; stems with pseudowhorls of leaves; leaves obovate, 5–33 × 1–8.5cm, base cuneate or attenuate, apex acuminate; petiole 1–4cm; inflorescence pendent, 5–50cm long; pedicels to 5mm, articulated above middle; flowers white (or with purple), 1.9–3cm; fruits orange or red; globose or depressed globose, 7–21mm diam.; seeds straw-coloured, hemispherical, 4–11 × 5–4 × 2–9mm. Forest; 620m.
**Distr.:** W Africa from Guinea to CAR, Congo (Kinshasa) and S to Angola and Zambia [tropical Africa].
**IUCN:** LC
**Bechati:** van der Burgt, 887 9/2006.

### *Dracaena fragrans* (L.) Ker-Gawl.
**Syn.** *Dracaena deisteliana* Engl.
Herb, to 3m, few-stemmed; stalk 1cm diam.; leaves sword-shaped, to 70 × 9cm; inflorescence terminal, erect; flowers white with pink lines, very fragrant. Forest; 1460m.
**Distr.:** tropical Africa [montane].

**IUCN:** LC
**Fosimondi:** Tchiengue, 2278 4/2005.

### *Dracaena phrynioides* Hook.
Herb, to 0.7m, short (to 0.1m) aerial stem; leaves 4–6; blades lanceolate, 18 × 7cm, spotted yellow; petiole c. 1mm wide; inflorescence terminal; peduncle 5cm; fruit bilobed, orange, sessile, head c. 2.5cm diam. Forest; 580–1370m.
**Distr.:** Liberia to Bioko, Rio Muni & Gabon [upper & lower Guinea (montane)].
**IUCN:** LC
**Bechati:** van der Burgt, 881 9/2006; **Fosimondi:** Tchiengue, 2209 4/2005.

## GRAMINEAE

### T. Cope (K) & B. Tchiengué (YA)

### *Digitaria pearsonii* Stapf
F.T.A. 9: 434 (1919); F.T.E.A. Gramineae: 643 (1982); Fl. Zamb. 10(3): 171 (1989).
Loosely caespitose perennial on a short branched rhizome; culms 40–70cm; nodes dark, glabrous; leaves 4–20 × 0.5–1.5cm, lanceolate to linear, flat, laxly hairy; inflorescence of 7–20 racemes, 4–12cm long, patent; pedicels 0.5–2.5mm, triangular; spikelets 1.8–2.5mm, lanceolate. Woodland & woodland clearings; 1800m.
**Distr.:** Burundi, Rwanda, Congo (Kinshasa), Tanzania, Uganda, Angola, Malawi, Mozambique & Zimbabwe [tropical Africa].
**IUCN:** LC
**Fosimondi:** Tchiengue, 2260 4/2005.

### *Eragrostis camerunensis* W.D.Clayton
Densely tufted perennial, 30–60cm; leaf-blades 3–7 × 0.1–0.2cm, usually rolled; panicle to 10cm; spikelets dark grey; pedicels 5–7 × 2mm. Grassland & fallow weed; 1800m.
**Distr.:** Cameroon & Nigeria [W Cameroon Uplands].
**IUCN:** NT
**Fosimondi:** Tchiengue, 2261 4/2005.

### *Leptaspis zeylanica* Nees
Agric. Univ. Wag. Papers 92(1): 39 (1992).
**Syn.** *Leptaspis cochleata* Thw.
Perennial, to 1m; stoloniferous, trailing, rooting at lower nodes, culms erect; leaf-blades oblong–oblanceolate, 10–35cm × 2.5–6cm, sharply pointed at tip, asymmetrical; sheaths longer than internodes, imbricate; inflorescence a panicle, 45–60cm; branches subverticillate, lower branches up to 20cm, 2–3 branches in a whorl, bearing short side branches with 1 male and 1 female spikelet, the male above; male spikelets 1–flowered, finely pubescent, the lemma 4mm long, the glumes half as long; stamens 6; female spikelets 4–6mm, 1–flowered, reddish, lemma inflated, conchiform, closed except for a small hole

near the apex, covered in hooked hairs. Undergrowth of dense, humid forest, edges of fields & roads; 1160m.
**Distr.:** tropical Africa & Asia [paleaotropics].
**IUCN:** LC
**Fosimondi:** Tchiengue, 2626 2/2006.

### *Melinis minutiflora* P.Beauv. var. *minutiflora*
Perennial, erect or ascending from a prostrate base, to 60cm; culms to 2m; leaves covered in sticky hairs and smelling strongly of molasses or linseed oil. Open hillsides.
**Distr.:** tropical Africa and introduced throughout the tropics [pantropical].
**IUCN:** LC
**Fosimondi:** Tchiengue, 2173 4/2005.

### *Panicum acrotrichum* Hook.f.
Straggling perennial, rooting from the nodes; leaf-blades lanceolate, 3–8 × 0.8–2.3cm, with transverse veins; panicle to 11cm, branched. Forest; 1430m.
**Distr.:** W Cameroon & Bioko [W Cameroon Uplands (montane)].
**Fosimondi:** Tchiengue, 2281 4/2005.

### *Paspalum conjugatum* Berg
Perennial; stoloniferous; culms erect, 20–80cm; leaf-blades linear to narrowly lanceolate, 4–20cm × 5–12mm, slightly ciliate at edges; ligules hairy; inflorescence 2 racemes, more or less joined, 5–15cm; short pedicels; spikelets orbicular, 1.5–1.7mm, yellow-green, closely appressed to the rachis of paired slender racemes; ciliate fringe from margins of upper glume, inner glume missing. Damp places in forest clearings, grassland & rough ground; 1430m.
**Distr.:** pantropical.
**IUCN:** LC
**Fosimondi:** Tchiengue, 2280 4/2005.

### *Poa annua* L.
Annual or short-lived perennial, tufted; culms to 2.5–30cm; leaves 1–14cm × 1–5mm, apex blunt; panicle 1–12cm, ovate or triangular; spikelets 3–10-flowered, 3–10mm long; lower glume 1.5–3mm long, 1-nerved; upper glume 2–4mm long, 3-nerved. Fallow, roadsides & grassland; 1800m.
**Distr.:** worldwide in temperate regions & tropical mountains [montane].
**IUCN:** LC
**Fosimondi:** Tchiengue, 1966 4/2004.

### *Sporobolus pyramidalis* P.Beauv.
**Syn.** *Sporobolus indicus* (L.) R.Br. var. *pyramidalis* (P.Beauv.) Veldkamp
Agric. Univ. Wag. Papers 92(1): 155 (1992).
Perennial, 30–160cm; caespitose in dense tufts, culms 2–5mm diam. at base; leaf-blades up to 50cm × 3–10mm, ribbon–like; basal sheaths narrow; inflorescence a panicle, 10–40cm, linear to pyramidal; branches spicate, those at base 5–10cm; spikelets not clustered,

1.7–2mm, elliptic, green, occasionally crimson, not acuminate; upper glume obtuse, $1/4$–$1/3$ length of spikelet. Savanna, often degraded or over–grazed, rough tracks, roadsides & villages.
**Distr.:** tropical & subtropical Africa.
**IUCN:** LC
**Fosimondi:** Tchiengue, 2171 4/2005.

## IRIDACEAE

### M. Cheek (K)

### *Gladiolus aequinoctialis* Herb.
Fl. Zamb. 12(4): 102 (1993); F.T.E.A. Iridaceae: 84 (1996); Goldblatt P., Gladiolus in Tropical Africa: 287 (1996).
**Syn.** *Acidanthera aequinoctialis* (Herb.) Baker
**Syn.** *Gladiolus aequinoctialis* Herb. var. *aequinoctialis*
Edwards's Bot. Reg. 28(Misc.): 85 (1842).
Herb, (40–)90–120cm; corm 2–3cm across; leaves 4–10, lower 2–5 basal, lanceolate, (0.6–)1.2–1.7cm across; spike (3–)5–8-flowered; bracts 5–8cm long; flowers white, showy, the lower 3 tepals streaked purple; perianth tube cylindric, (8.5–)12–14cm; tepals lanceolate, ± equal, 3.5–4 × 1.8–2.0cm; capsules 1.8–2.0cm long. Rocky places, often on wet ledges on steep cliffs & stony grassland.
**Distr.:** Sierra Leone, W Cameroon & Bioko [upper & lower Guinea].
**IUCN:** NT
**Fosimondi:** Tchiengue, 2169 4/2005; van der Waarde, 12 10/2004.
Note: here we follow Goldblatt (1996), who, in his mongraph of the tropical African species of *Gladiolus*, does not recognise varieties under this name.

## MARANTACEAE

### Y.B. Harvey (K)

Fl. Cameroun 4 (1965).

### *Hypselodelphys poggeana* (K.Schum.) Milne-Redh.
Lianescent herb, to several metres long; leaves linear-oblong, 8–15 × 3–9cm, abruptly acuminate, subtruncate; calloused portion of petiole above point of articulation 1–2cm; inflorescence loose, little-branched spikes, nearly straight, 5–9cm; internodes c. 5mm; bracts 2–3.5cm; flowers violet and white; fruit muricate, 3-lobed, 4.5–5.0 × 3.0–3.5cm; tubercles short and dense, not curved, less than 2mm. Lowland forest; 1300m.
**Distr.:** Sierra Leone to Cameroon, Bioko, Gabon, Congo (Brazzaville) & Congo (Kinshasa), Angola [Guineo-Congolian].
**IUCN:** LC
**Fosimondi:** Tah 321 4/2004.

### *Marantochloa monophylla* (K.Schum.) D'Orey
Bull. Jard. Bot. Nat. Belg. 65: 371 (1996).

**Syn.** *Marantochloa holostachya* (Baker) Hutch.
Rhizomatous herb; stems to 50cm; leaves solitary,
rarely 2, to 30 × 12cm, lanceolate, apex acuminate, base
subcordatae, lower surface violet, ± pubescent central
nerve; petiole to 20cm; callous swollen, 5–10mm;
inflorescence terminal, spike or dense panicle; peduncle
to 10–15mm; bracts persistent, to 15 × 4mm; pedicel to
4mm; corolla white, to 6–7mm; fruit pearly white, to
6mm diam., calyx persistent. Forest; 590m.
**Distr.:** SE Nigeria, Cameroon, Gabon, Congo
(Brazzaville) and Congo (Kinshasa) [Guineo-Congolian].
**IUCN:** LC
**Bechati:** Tchiengue, 2815 9/2006.
Note: Tchiengue 2815 is tentatively placed in this taxon,
it is a unicate, and not seen at Kew, Harvey, ii.2010.

# MUSACEAE

X. van der Burgt (K)

Fl. Cameroun 4 (1965).

## *Musa* spp.
Robust, tree-like herbs, to 3–7m; single-stemmed; with
basal suckers; stem leafless at maturity; leaves oblong,
to 2m; petiole distinct, at least 30cm; inflorescence
pendulous; bracts purple, 10–15 × 10–15cm; flowers in
lines in axils, females at base of inflorescence, males at
apex; fruits yellow, cylindric-angular, 10–30cm.
Cultivated from fruit, sometimes persisting for a few
years in abandoned farms; 320–400m.
**Distr.:** SE Asia, cultivated throughout tropics
[pantropical].
**IUCN:** LC
**Fosimondi:** Sight record, 10 9/2006.
**Uses:** FOOD — Infructescences.
Note: known only from habitat records (van der Burgt,
ix.2006), Harvey, ii.2010.

# ORCHIDACEAE

D. Roberts (K), C. Drinkell (K) & B.J. Pollard

Fl. Cameroun 34 (1988), 35 (2001) & 36 (2001).

## *Aërangis gravenreuthii* (Kraenzl.) Schltr.
Epiphyte; stem woody to 7 × 0.5cm; leaves distichous,
oblanceolate, strongly falcate, to 15 × 1.5–3cm,
narrowing basally; inflorescences pendent, axillary,
racemose, 10–20cm, 2–5-flowered; flowers white,
sometimes with an orange flush; dorsal sepal 20–32 ×
5–7mm; spur reddish, 4–8(–12)cm. Forest, woodland;
1800m.
**Distr.:** Bioko, Cameroon, Tanzania [lower Guinea &
Congolian (montane)].
**IUCN:** NT
**Fosimondi:** Simo 215 4/2004.

## *Ancistrorhynchus serratus* Summerh.
Epiphyte; stem 5–10cm; leaves 5–9, 5–18 × 0.7–1.2cm,
with parallel sides; apex irregularly serrate; inflorescence
< 1cm, several–multi-flowered; flowers white; labellum
2.5–3 × 2.5–4.5mm; spur 3.5–4.5mm. Forest; 1800m.
**Distr.:** SE Nigeria, W Cameroon, Bioko, Rio Muni, São
Tomé [lower Guinea].
**IUCN:** NT
**Fosimondi:** Simo 225 4/2004.

## *Angraecopsis ischnopus* (Schltr.) Schltr.
Epiphyte; stem to 5cm; leaves elliptic-oblong, elliptic-
ligulate or ligulate, 2–11 × 0.4–1.3cm; inflorescence
lax, 3–9cm, to 10-flowered; flowers small, resupinate,
dull greenish-white, fading brown; labellum to 8.5mm;
spur 1.2–3.8cm, very slightly or not at all swollen
apically. Forest, plantations; 1800m.
**Distr.:** Guinea, Sierra Leone, Nigeria, Cameroon [upper
& lower Guinea].
**IUCN:** LC
**Fosimondi:** Simo 234 4/2004; 236 4/2004.

## *Angraecum pungens* Schltr.
Epiphyte; stem pendent, to 50 × 0.2–0.3cm; leaves with
sharp points, not closely imbricate at base, oblong-
lanceolate, flattened, fleshy, 20–40 × 3–6mm; flowers
white; sepals 6–7mm long; labellum much broader than
long; spur 4.5mm long. Forest, riverine forest, swamp
forest; 1800m.
**Distr.:** SE Nigeria, Bioko, Cameroon, Congo
(Kinshasa) [lower Guinea & Congolian].
**IUCN:** LC
**Fosimondi:** Simo 216 4/2004.

## *Bulbophyllum cochleatum* Lindl. var. *cochleatum*
Epiphyte; pseudobulbs bifoliate, 0.8–7cm apart, 1.5–11
× 0.4–1.3cm; leaves 2.8–23 × 0.3–1.8cm; inflorescence
8–55cm; peduncle 5.2–43cm; rachis not thickened,
2.8–12cm, 14–64(–84)-flowered; sepals and petals
green, often stained purple-red or entirely so, with a
green base; labellum dark purple red, occasionally with
a yellow centre, with marginal hairs ≥ labellum width.
Forest; woodland; 1610–1800m.
**Distr.:** Guinea to Bioko, São Tomé, Cameroon, Gabon,
Sudan to E and S Africa [afromontane].
**IUCN:** LC
**Fosimondi:** Simo 217 4/2004; Tchiengue, 2574 2/2006.
Note: Tchiengue 2574 is tentatively placed in this taxon
(*B.* cf. *cochleatum* Lindl.) by Roberts, ii.2009.

## *Bulbophyllum cochleatum* Lindl. var. *tenuicaule*
(Lindl.) J.J.Verm.
Bull. Jard. Bot. Nat. Belg. 56: 230 (1986); Orchid
Monographs 2: 45 (1987).
**Syn.** *Bulbophyllum tenuicaule* Lindl.
Epiphytic herb; pseudobulbs narrowly conical or
cylindrical, 1–10 × 0.4–1.2cm, bifoliate, 1–9cm apart;
leaves 1.8–16 × 0.4–1.5cm; inflorescence 6.5–22cm;
peduncle 3–14.5cm; rachis slightly thickened, or not,
2–12cm, 8–60-flowered; flowers red-purple, whitish at
base; labellum dark brown-red or purple-red, with

marginal hairs ≥ labellum width; Forest, woodland; 1800m.
**Distr.:** SW Nigeria, W Cameroon, Bioko, São Tomé, Congo (Kinshasa), Rwanda, Uganda, Kenya [afromontane].
**IUCN:** LC
**Fosimondi:** <u>Simo 218</u> 4/2004.

### *Bulbophyllum falcipetalum* Lindl.
Epiphyte; pseudobulbs narrowly ovoid, 0.5 × 2cm, bifoliate; leaves narrowly elliptic, 4–6 × 23–26mm; rachis arching, to 3cm long; flowers yellow. Forest; 1800m.
**Distr.:** Ivory Coast, Ghana, Nigeria, Cameroon and Gabon [Guineo-Congolian].
**IUCN:** LC
**Fosimondi:** <u>Simo 224</u> 4/2004.

### *Bulbophyllum intertextum* Lindl.
Diminutive epiphyte; pseudobulbs 1-leafed, 0.2–2.5(–4)cm apart, 0.4–1.0 × 0.3–0.7cm; leaves elliptic to linear lanceolate, 0.7–10 × 0.3–1.1cm; inflorescence 2–30cm, 2–14(–20)-flowered; rachis arching, nodding, terete, usually zigzag bent, 0.2–19cm; floral bracts 1.5–4 × 1–2mm; flowers very pale yellowish or greenish, often suffused red. Lowland to submontane forest, forest patches in grassland; 1800m.
**Distr.:** tropical & subtropical Africa.
**IUCN:** LC
**Fosimondi:** <u>Simo 212</u> 4/2004.

### *Bulbophyllum josephi* (Kuntze) Summerh. var. *josephi*
Epiphytic or epilithic herb; pseudobulbs unifoliate, 0.7–3cm apart, 1.5–4 × 0.6–2.4cm; leaf-petiole 2–35mm; lamina lanceolate, 4.5–28 × 0.9–3.2cm, slightly emarginate; inflorescence 8.5–40cm, 7–80-flowered; rachis arching to pendulous; flowers cream, tinged pink, pale green or yellowish; column without teeth along the adaxial margins. Forest; 1680–1800m.
**Distr.:** Cameroon, Congo (Kinshasa), Rwanda, Burundi, Ethiopia, Kenya, Tanzania, Malawi, Mozambique [afromontane].
**IUCN:** LC
**Fosimondi:** <u>Simo 223</u> 4/2004; <u>235</u> 4/2004; <u>Tchiengue, 2580</u> 2/2006.

### *Bulbophyllum maximum* (Lindl.) Rchb.f.
**Syn.** *Bulbophyllum oxypterum* (Lindl.) Rchb.f.
Epiphyte; pseudobulbs 2(–3)-leafed, 2–10cm apart, 3.5–10 × 1–3cm; leaves oblong to linear-lanceolate, maximum width usually just above middle, 3.8–20 × 1.3–5cm; inflorescence 15–90cm; 16–120-flowered; rachis bladelike, 6–56 × 8–50mm; floral bracts spreading to reflexed, 2.5–7 × 2–4mm; flowers yellowish or greenish, spotted purple. Lowland primary and secondary forest, montane forest, savanna wooodland; 1800m.
**Distr.:** tropical & subtropical Africa.
**IUCN:** LC
**Fosimondi:** <u>Simo 220</u> 4/2004.

### *Bulbophyllum nigericum* Summerh.
Epiphytic or epilithic herb; pseudobulbs bifoliate, ovoid, 0.8–2.5cm apart, 1.3–2.7 × 0.7–1.5cm, obtusely 4-angled, drying bright yellow; petiole 1–2mm; lamina lanceolate to linear-lanceolate, 3–7 × 0.5–1.2cm, ± emarginate; midrib prominent abaxially; inflorescence 8–23cm, 7–30-flowered; peduncle erect, 4–9.5cm × 3.5mm, elliptic in section, with 9–14 tubular scales, the longest 7–12mm; rachis 4-angled in section, ± zigzag bent, 3.5–14 × 0.4cm; floral bracts conspicuous, spreading, concave, ovate, 7–12 × 4–6.5mm, fibrous; flowers distichous, many open simultaneously; sepals very pale green or yellow, occasionally stained purple; petals purplish, spotted deep purple or yellowish; labellum bright yellow, suffused purplish-brown basally, or orange. Forest; 1800m.
**Distr.:** S Nigeria, W Cameroon [W Cameroon Uplands].
**IUCN:** VU
**Fosimondi:** <u>Simo 229</u> 4/2004.

### *Bulbophyllum oreonastes* Rchb.f.
**Syn.** *Bulbophyllum zenkerianum* Kraenzl.
Epiphytic or epilithic herb; pseudobulbs bifoliate, 0.7–4.0cm apart, 0.4–3.5 × 0.4–1.2cm; leaves elliptic to linear-lanceolate, 0.6–8.2 × 0.4–2.0cm; inflorescence 1.5–17.5cm, 5–36-flowered; floral bracts yellow, often suffused red; tepals yellow, orange to dark red-purple, with conspicuous dark longitudinal stripes; labellum 1.4–2.5 × 0.6–1.2mm. Forest; 1610–1800m.
**Distr.:** tropical and subtropical Africa [afromontane].
**IUCN:** LC
**Fosimondi:** <u>Simo 222</u> 4/2004; <u>Tchiengue, 2286</u> 4/2005; <u>2572</u> 2/2006.

### *Bulbophyllum sandersonii* (Hook.f.) Rchb.f. subsp. *sandersonii*
F.T.E.A. Orchidaceae: 320 (1984); Orchid Monographs 2: 108 (1987); Fl. Rwanda 4: 533 (1988); Fl. Cameroun 35, in press (1998).
**Syn.** *Bulbophyllum tentaculigerum* Rchb.f.
Epiphyte; pseudobulbs bifoliate, 1.2–6.5cm apart, narrowly ovoid; leaves lanceolate to linear, 3.5–26 × 0.5–2.5cm, oblique; inflorescence 5.5–30cm; rachis 1.5–9 × 0.3–1.1cm; floral bracts distinctly narrower than the fully developed part of the rachis; flowers usually placed along an excentric line on the rachis, 3–10mm apart; flowers yellowish or greenish suffused purple. Lowland to montane forest; 1800m.
**Distr.:** Cameroon, Gabon, Congo (Kinshasa), E Africa, Zambia, Malawi, Zimbabwe, Mozambique, South Africa [tropical & subtropical Africa].
**IUCN:** LC
**Fosimondi:** <u>Simo 208</u> 4/2004.

### *Bulbophyllum sandersonii* (Hook.f.) Rchb.f. subsp. *stenopetalum* (Kraenzl.) J.J.Verm.
Bull. Jard. Bot. Nat. Belg. 56: 234 (1986); Orchid

Monographs 2: 110 (1987).
Epiphyte; pseudobulbs bifoliate, 1.2–6.5cm apart, narrowly ellipsoid to narrowly ovoid; leaves narrowly lanceolate to linear-lanceolate, 3.5–26 × 0.5–2.5cm, oblique; inflorescence 5.5–30cm; rachis 6–14.5 × 0.2–0.5cm; floral bracts as wide as the fully developed part of the rachis; flowers usually placed along the median line of the rachis, 5.5–11mm apart, yellowish or greenish suffused purple; labellum yellowish with purple dots. Lowland to lower montane forest, occasionally in secondary forest; 1350m.
**Distr.:** Liberia, Ivory Coast, Ghana, Nigeria, Cameroon, Gabon, Congo (Kinshasa) [Guineo-Congolian].
**IUCN:** LC
**Fosimondi:** Tchiengue, 2243 4/2005.

*Calanthe sylvatica* (Thouars) Lindl.
F.T.E.A. Orchidaceae: 282 (1984).
**Syn.** *Calanthe corymbosa* Lindl.
Terrestrial herb, to 70cm; pseudobulbs conical, 2–5 × 1.5cm; leaves ± rosulate, suberect, to 35 × 12cm, lanceolate; petiole 5–25cm; peduncle ± with 2 sheath-like leaves; inflorescence many flowered; flowers white or purple; petals 8–25 × 4–14mm, elliptic or oblanceolate; lip 11–15 × 6–25mm; spur slender, 1–4cm long. Forest; 1800m.
**Distr.:** tropical and subtropical Africa [afromontane].
**IUCN:** LC
**Fosimondi:** Simo 230 4/2004.

*Cyrtorchis arcuata* (Lindl.) Schltr. var. *variabilis* Summerh.
Fl. du Cameroun 36(3): 726 (2001).
Epiphytic herb; stem woody, to 30cm, old leaf bases persistent; leaves coriaceous, fleshy, 7–20 × 1–3.8cm, linear; inflorescences to 18cm, to 10-flowered; pedicel and ovary 15–50mm long; bracts 10–30 × 10–16mm; flowers stellate, white, maturing pale orange; petals 8–38 × 4–5mm; lip 8–42 × 3.5–6mm, lanceolate, acuminate; spur 3–10cm long. On fallen branches 1350–1800m.
**Distr.:** Senegal, Guinea (Conakry), Liberia, Ivory Coast, Ghana, Togo, Benin, Nigeria, Cameroon, CAR, Congo (Kinshasa), Rwanda, Burundi, Uganda, Kenya, Tanzania, Zambia, Malawi, Zimbabwe, Mozambique [tropical Africa].
**IUCN:** LC
**Fosimondi:** Simo 211 4/2004; Tchiengue, 2240a 4/2005.
Note: Simo 211 and Tchiengue 2240a are both sterile and have tentatively been placed in this taxon, Harvey, viii.2009.

*Cyrtorchis brownii* (Rolfe) Schltr.
Epiphyte, to 8–20(–40)cm; stem robust, branched; leaves 4–9, 3–12 × 0.7–1.5cm, linear or oblong-elliptic, unequally bilobed at the apex; inflorescence very dense, 3–9cm, 8–20-flowered; flowers small, non-resupinate,

white, fragrant; labellum 5.5–10.5 × 2.5–3mm; spur 2–3cm. Evergreen forest; 1610m.
**Distr.:** Sierra Leone, Cameroon, Gabon, CAR, Congo (Brazzaville), Congo (Kinshasa), Uganda, Tanzania, Malawi [tropical Africa].
**IUCN:** LC
**Fosimondi:** Tchiengue, 2573 2/2006.

*Cyrtorchis letouzeyi* Szlach. & Olszewski
Fl. Cameroun 36: 722 (2001).
Herb, to 10–15cm; leaves 7–12, 16.5 × 2.8cm, base imbricate, linear or oblong elliptic, apex (obliquely) emarginate; inflorescence 4–13cm, with 4–10 flowers; flowers white to pale pink-orange; bracts to 10mm, elliptic, glabrous; pedicel and ovary to 28mm; petals 21 × 5mm, oblong lanceolate; spur to 85–90 × 4mm, tapering to point at apex. Epiphyte of submontane forest; 1800m.
**Distr.:** Cameroon, Gabon, CAR [Guineo-Congolian].
**IUCN:** EN
**Fosimondi:** Simo 232 4/2004.
Note: Simo 232 is tentatively placed in this taxon by Roberts, xii.2008 (*Cyrtorchis* cf. *letouzeyi*).

*Cyrtorchis ringens* (Rchb.f.) Summerh.
Epiphyte, stem arcuate, woody, to 30cm in old plants; leaves linear, usually 6–7, thick and leathery, 7.5–12.5 × 1.2–2.5cm; inflorescence 6–7(–16)cm, to 16-flowered; flowers closely placed, creamy white, sweetly scented; spur to 3cm. Upland evergreen forest; 1800m.
**Distr.:** Senegal, Sierra Leone to Cameroon, São Tomé, Congo (Kinshasa), Burundi, Uganda, Tanzania, Zambia, Malawi, Zimbabwe [afromontane].
**IUCN:** LC
**Fosimondi:** Simo 210 4/2004.

*Cyrtorchis* sp. of Fosimondi
Epiphyte, to 25cm; stem arcuate; leaves 11, 9.5–13 × 1.8–2.3cm, linear-lanceolate, thick and leathery; inflorescence to 8cm, to 8-flowered; flowers not present. Epiphyte of submontane forest; 1350m.
**Distr.:** Cameroon (Fosimondi).
**Fosimondi:** Tchiengue, 2239 4/2005.

*Diaphananthe bueae* (Schltr.) Schltr.
Epiphyte; stem 1.5cm long, 5mm diam.; leaves ± 5, linear, unequally bilobed at apex, 4–7(–19) × 0.5–1.5cm, twisted at base; inflorescences 5–7cm long, laxly many-flowered; peduncle 6–12mm long; flowers pale yellow; pedicel and ovary 5–8mm long; petals linear to obovate, 5–8 × 1.3–2.2mm; spur swollen and slightly lobed at base. Forest; 1800m.
**Distr.:** Cameroon, Ivory Coast, Uganda [lower Guinea (montane)].
**IUCN:** EN
**Fosimondi:** Simo 228 4/2004.
Note: Simo 228 is sterile.

*Eulophia horsfallii* (Batem.) Summerh.
Robust terrestrial herb, 1–3m; leaves 3–5, lanceolate to
oblanceolate, 30–200 × 5–15cm, ribbed; scapes to 3m;
inflorescence a long raceme, laxly 5–50-flowered;
flowers large, fleshy, purple with maroon sepals and
petals; labellum 2–4 × 1.4–4cm. Forest edge, swamp,
fallow; 1800m.
**Distr.:** tropical and subtropical Africa.
**IUCN:** LC
**Fosimondi:** Tchiengue, 1931 4/2004.

*Habenaria weileriana* Schltr.
F.W.T.A. 3: 194 (1968).
Epilithic, occasionally epiphytic, or rarely terrestrial
herb, to 40cm, often occurring in large colonies; leaves
linear or narrowly lanceolate, 8–10 cauline, 6.5–11.5 ×
0.7–1cm; inflorescence lax, 2.7–6cm, to (2–)3(–4)-
flowered; flowers white; labellum trilobed; median lobe
8–9 × 2–2.5mm; spur 4–5.5cm. Forest and stream
banks, rocks in rivers and streams; 250m.
**Distr.:** SE Nigeria, W Cameroon, Gabon [lower
Guinea].
**IUCN:** NT
**Bechati:** van der Burgt, 852 9/2006.

*Liparis nervosa* (Thunb.) Lindl. var. *nervosa*
Gen. et Sp. Orch.: 26 (1830); Fl. Afr. Cent. Orchidaceae:
275 (1984); F.T.E.A. Orchidaceae: 298 (1984).
**Syn.** *Liparis guineensis* Lindl.
**Syn.** *Liparis rufina* (Ridl.) Rchb. f. ex Rolfe
Terrestrial, epilithic or rarely epiphytic herb, to 70cm;
stem basally swollen; leaves 2–5, petiolate, sheathing,
lanceolate, to 35 × 7.5cm; peduncle to 55cm; rachis
many-flowered, to 15cm; flowers green or yellow to
reddish or purplish-brown. Forest, marshy grassland, or
stony grassland and on wet rock outcrops; 250m.
**Distr.:** tropics and subtropics of the World, including
Africa, India to Japan, Phillipines, Costa Rica, W
Indies, S America [pantropical].
**IUCN:** LC
**Bechati:** van der Burgt, 853 9/2006.
Note: van der Burgt 853 is tentatively placed in this
taxon by Cribb, ii.2009.

*Manniella gustavi* Rchb.f.
Terrestrial herb, 50–90cm with stout fleshy roots; stem
very short; leaves radical, membranous, 4–7; petiole to
15cm; blades obliquely ovate, to 16 × 7cm, usually
white-spotted; inflorescence glandular-pubescent; scape
to 45cm, raceme to 40cm; floral bracts pubescent-
papillose with ciliate margins; flowers brownish-pink;
lateral sepals revolute, convex; labellum 2–3 × 3mm.
Forest; 1160m.
**Distr.:** Sierra Leone to Bioko, São Tomé, Cameroon,
Congo (Brazzaville), Congo (Kinshasa), Uganda,
Tanzania [afromontane].
**IUCN:** LC
**Fosimondi:** Tchiengue, 2625 2/2006.

*Polystachya adansoniae* Rchb.f. var. *adansoniae*
Epiphyte or epilith, to 20cm; pseudobulbs nearly
cylindric-conical, 2.5–9cm, 2–4-leafed; leaves linear,
narrowly oblong or linear-ligulate, 8–19 × 0.6–1.3cm;
inflorescence racemose, 5–12(–20)cm, many-flowered;
flowers non-resupinate, white to greenish-yellow; petals
and labellum purple; labellum 1.8–2.8 × 1.1–2.8mm;
spur 1.5–2.5mm, sacciform. On trees, or rocks, in
forest; 1350m.
**Distr.:** Guinea to Cameroon, Gabon, Congo (Kinshasa),
E Africa, Zambia, Malawi, Angola, Zimbabwe [tropical
& subtropical Africa].
**IUCN:** LC
**Fosimondi:** Tchiengue, 2240b 4/2005.

*Polystachya bicalcarata* Kraenzl.
A small densely tufted epiphyte, to 20cm; pseudobulbs
1.5–7.5 × 0.05–0.2cm, unifoliate; leaves linear, fleshy,
4–16 × 0.2–0.4cm, articulated 2 to 3mm above the apex
of the pseudobulb; inflorescence paniculate, 2.5–12cm,
3(–11)-flowered, very slender, axis shortly branching
flowers non-resupinate, borne on a tight head on each
branch, white, purple and green; mentum bifid;
labellum 5–6 × 3–4mm. Forest; 1800m.
**Distr.:** Nigeria, W Cameroon, Bioko [W Cameroon
Uplands].
**IUCN:** VU
**Fosimondi:** Simo 226 4/2004.

*Polystachya bifida* Lindl.
Epiphyte, 20–60cm; stems often pendulous, 20–40cm,
fusiform, 4–12-leaved; leaves linear-lanceolate, 4–21 ×
0.4–1.5cm; inflorescence lax, unbranched, 4–20cm;
3–25-flowered; peduncle and rachis glabrous; flowers
non-resupinate, white; dorsal sepal purple; labellum 7–8
× 3.5–4mm, ± rhombiform; spur to 7mm, conical-
sacciform. Forest, marshy forest; 1450–1610m.
**Distr.:** SE Nigeria, Cameroon, Bioko, São Tomé, Gabon,
Congo (Kinshasa), Rwanda [lower Guinea & Congolian].
**IUCN:** LC
**Fosimondi:** Tchiengue, 2194 4/2005; 2571 2/2006.
Note: Tchiengue 2194 is tentatively placed in this taxon
by Roberts & Drinkell, vii.2007.

*Polystachya cultriformis* (Thouars) Spreng.
Epiphyte 6–40cm; pseudobulbs unifoliate, narrowly
cylindric to conical, 1.4–18 × 0.2–1.2cm; leaf obovate,
ovate or elliptic, auriculate at base, 3.2–36 × 1.2–5.5cm,
± undulate; inflorescence terminal, branched,
4.4–29cm, bearing up to 60 flowers successively;
flowers non-resupinate, very variable in size and
colour; labellum 4–7.8 × 3–6mm; spur 4–7mm,
sacciform. Forest; 1350–1800m.
**Distr.:** Ivory Coast, Cameroon, Bioko, Gabon, Congo
(Kinshasa), Burundi, E Africa to S Africa, Madagascar,
Mascarene Is., Seychelles [tropical & subtropical Africa].
**IUCN:** LC
**Fosimondi:** Simo 221 4/2004; Tchiengue, 2238 4/2005.

*Polystachya fusiformis* (Thouars) Lindl.

Suberect or pendent epiphyte, or epilith, to ± 60cm; stems (pseudobulbs) cylindrical to fusiform, superposed, longitudinally ridged, drying yellow, to 22 × 0.3–0.4cm; leaves 3–7, oblong-lanceolate, lanceolate to oblanceolate, 5–16 × 0.6–1.6(–3.2)cm, largest borne apically; inflorescence terminal, paniculate, 3–8(–15)cm, densely 20–80-flowered; peduncle pubescent; flowers miniscule, non-resupinate, persistent on developed ovary, cream, yellow-green tinged or entirely purple or mauve; pedicel and ovary 4mm, glabrous; labellum 2–2.5 × 2.5mm; spur 1mm, sacciform. Forest; 1800m.
**Distr.:** Ghana, Cameroon, Bioko, Congo (Kinshasa), Rwanda, Burundi, E Africa, Zambia, Malawi, Zimbabwe, South Africa, Mascarene Is. [afromontane].
**IUCN:** LC
**Fosimondi:** Simo 214 4/2004.

*Polystachya polychaete* Kraenzl.

Stout epiphyte, 15–50cm; pseudobulbs 8–10 × 1cm, 3–7-leafed; leaves 3–7, ligulate, (6–)12–18(–30) × 0.8–2.5cm, apex conspicuously bilobed; inflorescence a dense unbranched terminal raceme, 10–26cm, densely many-flowered; rachis ciliate; flowers miniscule, non-resupinate, yellow, greenish-yellow, white, greenish-white or cream; labellum 1.2–2.2 × 1.5–2.2mm; spur 1.3–2mm, sacciform-conical. Forest; 1800m.
**Distr.:** Sierra Leone to Cameroon, Bioko, Gabon, Congo (Brazzaville) & Congo (Kinshasa), Rwanda, Uganda, E Africa [Guineo-Congolian].
**IUCN:** LC
**Fosimondi:** Simo 213 4/2004.

*Polystachya rhodoptera* Rchb.f.

Epiphyte, to 50cm; without pseudobulbs; leaves 3–9, 5–20 × 0.3–2.5cm, lanceolate to oblong, acute; inflorescence dense, to 10.5cm, 8–40-flowered, simple or 1–2-branched; branches very short; peduncle and rachis densely pubescent; flowers non-resupinate, white or yellow, often pink-tinged; labellum 5–6.6 × 4–5.5mm; spur 4–5mm, conical. Forest, shaded branches above water; 1800m.
**Distr.:** Sierra Leone to Cameroon, Bioko, Rio Muni, CAR, Príncipe, São Tomé, Gabon, Congo (Kinshasa) [Guineo-Congolian].
**IUCN:** LC
**Fosimondi:** Simo 219 4/2004.

*Polystachya tessellata* Lindl.

**Syn.** *Polystachya concreta* (Jacq.) Garay & H.R.Sweet Kew Bull. 42: 724 (1987); Orquideologia 9: 206 (1974).
Epiphyte, (10–)20–60cm; pseudobulbs 15 × 0.5–0.7cm, 3–5-foliate; leaves oblanceolate or elliptic, (3–)10–30 × 0.8–6.0cm; inflorescence paniculate, 10–50cm; branches secund, distant, densely 20–200-flowered; rachis and peduncle covered in sheaths; flowers small, non-resupinate, cream, yellow, clear green or red-purple. Forest, savanna or woodland; 1800m.

**Distr.:** tropical and subtropical Africa.
**IUCN:** LC
**Fosimondi:** Simo 233 4/2004.

*Tridactyle anthomaniaca* (Rchb.f.) Summerh.

Epiphyte; stems semi-pendent to 2m × 0.5–0.6cm; leaves numerous, fleshy-coriaceous, linear to narrowly elliptic-oblong, 3.5–11 × 0.6–1.9cm, shiny above, matt below; inflorescence very short, to 1cm, 2–4-flowered; flowers small, resupinate, green, pale green, yellow or white; labellum 3–6 × 1–2mm; spur filiform, (6–)11–16mm. Riverine, swamp and lower montane forest, plantations (coffee, cocoa, guava), often above water, or in sunny positions; 1800m.
**Distr.:** Sierra Leone to Congo (Kinshasa), CAR, E Africa, S to Mozambique, Malawi, Zambia, Zimbabwe [afromontane].
**IUCN:** LC
**Fosimondi:** Simo 209 4/2004.

# PALMAE

## B. Tchiengué (YA)

*Elaeis guineensis* Jacq.

Single-stemmed tree, to 20m; leaves crowded, pinnately-compound, to 5m, arching, basal leaflets modified as spines; inflorescences partially hidden at leaf bases; fruits oblong-globose, angular by mutual compression, c. 4 × 3cm, ripening orange-red, marked brown. Plantations & farmbush, forest; 320–680m.
**Distr.:** tropical Africa, but cultivated throughout the tropics .
**IUCN:** LC
**Bechati:** Sight record, 12 9/2006.
**Uses:** FOOD ADDITIVES — Infructescences: oils/fats.
Note: known only from habitat notes and images of oil production (Tchiengue ix.2006), Harvey, ii.2010.

# ZINGIBERACEAE

## Y.B. Harvey (K) & D. Harris (E)

Fl. Cameroun 4 (1965).

*Aframomum flavum* Lock

Bull. Jard. Bot. Nat. Belg. 48: 393 (1978).
**Syn.** *Aframomum hanburyi sensu* Koechlin
Fl. Cameroun 4: 65 (1965).
Herb, to 4m; leaves narrowly elliptic, c. 45 × 8–11cm, acuminate, base cuneate, glabrous, ligule 5–8mm, suborbicular; inflorescence 4–6-flowered, peduncle 4–6(–20)cm, bracts broadly ovate, coriaceous, puberulent, 4.5 × 3.5cm; flowers yellow; fruit smooth, red, calyx persistent. Forest, clearings & thicket; 470–1120m.
**Distr.:** Cameroon & Rio Muni [lower Guinea].
**IUCN:** pending
**Bechati:** van der Burgt, 860 9/2006; **Fosimondi:** Tchiengue, 2644 2/2006.

*Aframomum zambesiacum* (Baker) K.Schum.
**Syn.** *Aframomum chlamydanthum* Loes. & Mildbr.
Herb, to 3m; leaves narrowly elliptic, base cuneate,
margins and midrib of lower leaf hairy, ligule short,
rounded; inflorescence capitate, >20-flowered; flowers
whitish, labellum with purple centre; fruits deeply
grooved, thick walled, seeds shiny, dark brown, rough.
Forest; 1800m.
**Distr.:** Nigeria, Cameroon, Congo (Kinshasa) & E
Africa to Malawi [afromontane].
**IUCN:** LC
**Fosimondi:** Tchiengue, 2249 4/2005.

### *Aframomum* sp. 1 of Fosimondi
Herb, to 2.5m; leaves narrowly elliptic, c. 2–3.5 ×
20cm, acuminate, base cuneate, glabrous, veins not
prominent; inflorescence bracts broadly ovate (to
2.5cm), coriaceous, minute hairs at apex of external
face; flowers white, 5.5cm long. Submontane forest;
1500m.
**Distr.:** Cameroon (Fosimondi).
**Fosimondi:** Tchiengue, 2272 4/2005.

### *Aframomum* sp. 2 of Fosimondi
Herb, to 2m; leaves narrowly lanceolate, c. 3–3.5 ×
18–20cm, acuminate, base attenuate, glabrous, veins
prominent on lower surface of dried material;
inflorescence bracts ovate, 1.5–2cm wide, glabrous,
coriaceous; flowers purple, c. 7cm long. Submontane

forest; 1220–1360m.
**Distr.:** Cameroon (Fosimondi).
**Fosimondi:** Tchiengue, 2598 2/2006; 2645 2/2006.

### *Aframomum* sp. 3 of Fosimondi
Herb, to 2.5m; leaves narrowly elliptic, c. 2–3 ×
23–28cm, acuminate, base cuneate, glabrous, veins not
prominent, with a distinct false petioles; inflorescence
bracts deep purple, ovate, coriaceous, glabrous, 15mm
wide; corolla tube with purple exterior, yellow-white
interior. Submontane forest; 1450m.
**Distr.:** Cameroon (Fosimondi).
**Fosimondi:** Tchiengue, 2186 4/2005.

### *Renealmia africana* (K.Schum.) Benth.
Herb, from short thick rhizome; leaves with a distinct
false petiole, blades elliptic, to 30 × 8cm or more, apex
acuminate, base narrowly cuneate, veins prominent on
both surfaces; inflorescence arising from the rhizome,
near the leaves; flowers small, delicate, whitish-
translucent, lateral inflorescence branches spreading to
upright; fruits spherical to ellipsoid, c. 8mm diam.,
reddish becoming black. Forest; 1300–1700m.
**Distr.:** Nigeria to Congo (Kinshasa) [lower Guinea &
Congolian (montane)].
**IUCN:** LC
**Fosimondi:** Tah 308 4/2004; Tchiengue, 1940 4/2004.
Note: Tah 308 & Tchiengue 1940 are tentatively placed
in this taxon, Harvey, i.2010.

# GYMNOSPERMAE
## PINOPSIDA

### CUPRESSACEAE

B. Tchiengué (YA)

***Cupressus lusitanica*** Mill.
Dallimore & Jackson, Handbook of Coniferae: 206 (1923).
Tree, to 30–35m, evergreen, monoecious; trunk monopodial, large trees buttressed, up to 2m dbh; leaves scale-like; seed cones solitary or in groups, near the upper ends of lateral branches, terminal on short, leafy branchlets, maturing in 2 growing seasons, persistent. Cultivated.
**Distr.:** Mexico to Honduras, introduced into tropical and subtropical Africa, S America and S Asia [pantropical and temperate regions].
**IUCN:** LC
**Fosimondi:** Photographic record, 4 4/2005.
Note: known only from photographic record, Harvey, ii.2010.


### PODOCARPACEAE

B. Tchiengué (YA)

***Podocarpus milanjianus*** Rendle
Dioecious shrub or tree, to 35m; bark exfoliating in papery flakes; slash pale brown; stems much-branched, sympodial; leaves spreading, alternate, linear-lanceolate, (5–)10–15 × 0.5–1.5cm, stomata on lower side only, midrib prominent and raised below; male cones solitary or paired, flesh-pink, c. 3cm; female cones solitary; fruit green, obovoid to subglobose, c. 1cm long; receptacle well-developed, obconical to subglobose, fleshy, red; seeds 1–2, subglobose, 8–9mm. Forest; 1920m.
**Distr.:** Cameroon, Congo (Kinshasa), Angola, Sudan to Zimbabwe [afromontane].
**IUCN:** LC
**Fosimondi:** Tchiengue, 2646 2/2006.

# PTERIDOPHYTA

Fl. Cameroun 3 (1964)
Note: here, we are not currently following Roux, J.P. (2009). Synopsis of the Lycopodiophyta and Pteridophyta of Africa, Madagascar and neighbouring islands. Strelitzia 23: 1–296

# LYCOPSIDA
Y.B. Harvey (K)

## SELAGINELLACEAE

*Selaginella vogelii* Spring
Terrestrial, in deep shade; stolons long-creeping, pink; stems erect, at least partly pink when dry, pubescent; frond-like portion triangular, 15–40cm long and wide; branches pubescent on underside; lateral microphylls oblong-lanceolate, with entire margins; median leaves with basal auricles, long acuminate-aristate. Forest.
**Distr.:** tropical Africa.
**IUCN:** LC
**Fosimondi:** Photographic record, 5 4/2005.
Note: no collection made, tentative determination from photograph, Harvey, ii.2010.

# FILICOPSIDA
P.J. Edwards (K) & B. Tchiengué (YA)

## ASPLENIACEAE

*Asplenium paucijugum* Ballard
Bull. Jard. Bot. Nat. Belg. 55: 147 (1985); Acta Botanica Barcinonensia 40: 8 (1991).
**Syn.** *Asplenium variabile* Hook. var. *paucijugum* (Ballard)
Terrestrial, low epiphyte or on rocks; rhizome creeping, 3–6mm diam.; fronds tufted, erect, 20–60cm long, ± 1-pinnate; stipe pale brown or green; lamina dark green with 1–2 pairs of subopposite pinnae; sori closely parallel long the oblique veins, 7–29mm long. Forest, sometimes in stream beds; 1370m.
**Distr.:** tropical Africa.
**IUCN:** LC
**Fosimondi:** Tchiengue, 2217 4/2005.

## CYATHEACEAE

*Cyathea camerooniana* Hook. var. *camerooniana*
**Syn.** *Alsophila camerooniana* (Hook.) R.M.Tryon var. *camerooniana*
Acta Botanica Barcinonensia 31: 26 (1978).
Tree fern, to 4m; trunk slender, 0.6–3m tall, not spiny; fronds 1.2 × 0.4 to 2.5 × 0.55m; stipe c. 10cm; pinnae sessile, gradually reducing in size in lower $^1/_4$ of frond;

sori near costules, at forking of nerve. Forest & streambanks; 1370m.
**Distr.:** Guinea to Bioko, Cameroon, Congo (Brazzaville) & Gabon [Guineo-Congolian (montane)].
**IUCN:** LC
**Fosimondi:** Tchiengue, 2225 4/2005.

*Cyathea manniana* Hook.
**Syn.** *Alsophila manniana* (Hook.) R.M.Tryon
Acta Botanica Barcinonensia 31: 27 (1978).
Tree fern, to 10m, erect, slender; stipe sharply spinose; fronds arching to horizontal, to 2.5 × 1m, 3-pinnate, old fronds not persistent. Forest & farmbush; 1400m.
**Distr.:** tropical Africa [afromontane].
**IUCN:** LC
**Fosimondi:** Tchiengue, 2224 4/2005.

## DENNSTAEDTIACEAE

*Pteridium aquilinum* (L.) Kuhn subsp. *aquilinum*
Terrestrial fern; thicket forming; rhizome long-creeping, subterranean; fronds to 1.5m tall; stipe erect, the base black (remainder brown); lamina 3-pinnate pinnatisect; sori marginal with fimbriate indusia on both sides. Grassland and forest; 1610–1680m.
**Distr.:** cosmopolitan [montane].
**IUCN:** LC
**Fosimondi:** Sight record, 6 4/2005, 2/2006.

## MARATTIACEAE

*Marattia fraxinea* J.Sm. var. *fraxinea*
Very large terrestrial fern; rhizome erect, to 40 × 30cm; fronds tufted to 4m, stiff, fleshy; stipe with brown flushing and long white- or green-streaks; swollen base with a pair of green to dark brown, thick, fleshy stipules; lamina ovate in outline, 2-pinnate, to 2 × 1m. Forest; 400–1500m.
**Distr.:** palaeotropical [montane].
**IUCN:** LC
**Bechati:** van der Burgt, 865 9/2006; **Fosimondi:** Tchiengue, 2268 4/2005.

## OSMUNDACEAE

*Osmunda regalis* L. var. *regalis*
Terrestrial fern; rootstock erect becoming very large, embedded in matted, fibrous roots; fronds erect to 2m; lamina 2-pinnate to 1m long; 2–5 fertile pinnae borne in the apical portion of some fronds. Near streams; 1800m.
**Distr.:** subsaharan Africa, western Indian Ocean region, temperate Europe, Asia & America [pantropical].
**IUCN:** LC
**Fosimondi:** Tchiengue, 1971 4/2004.

# POLYPODIACEAE

*Lepisorus excavatus* (Bory ex Willd.) Ching
Zink M., Systematics of *Lepisorus*: 37 (1993).
**Syn.** *Pleopeltis excavata* (Bory ex Willd.) Moore
Acta Botanica Barcinonensia 33: 16 (1982).
Epiphytic fern; rhizome long-creeping; stipe short;
lamina very thin, glabrous, 15–30 × 2–3.5cm; sori
circular, large, in a single series each side of the midrib.
Forest-grassland transition; 1350m.
**Distr.:** Guinea to Bioko & Cameroon [Guineo-
Congolian (montane)].
**IUCN:** LC
**Fosimondi:** Tchiengue, 2241 4/2005.

*Loxogramme abyssinica* (Baker) M.G.Price
Amer. Fern. J. 74(2): 61 (1984).
**Syn.** *Loxogramme lanceolata* (Sw.) C.Presl
Epiphytic fern; rhizome wide-creeping, 1–2mm diam.;
fronds mostly with short stripes; lamina entire,
glabrous, very leathery, 5–30 × 1–2.5cm; sori very
elongated, forming oblique parallel lines near midrib.
Forest-grassland transition; 540–1150m.
**Distr.:** tropical Africa [afromontane].
**IUCN:** LC
**Bechati:** Tchiengue, 2834 9/2006; **Fosimondi:**
Tchiengue, 2636 2/2006.

*Platycerium stemaria* (P.Beauv.) Desv.
Epiphyte; rhizome scales light brown; basal fronds
erect, cuneate, to 60 × 30cm, truncate at apex, white
stellate-pubescent beneath; fertile fronds to 90cm, once
to twice dichotomously forked, apex irregularly dentate,
velvety white stellate-pubescent beneath. Forest and
plantations; 580m.
**Distr.:** Senegal to Bioko, Príncipe, São Tomé,
Cameroon, CAR, Gabon, Congo (Brazzaville), Congo
(Kinshasa), Angola, Sudan and Comoros [Guineo-
Congolian].
**IUCN:** LC
**Bechati:** Fosimondi 6 9/2006.
Note: photographic record only, see Plate 1C. Tentative
determination, Harvey, iii.2010.

# INDEX

Accepted names in roman, synonyms in *italic*. **Bold** text indicates figures.

Achyrospermum oblongifolium  124
*Acidanthera aequinoctialis*  154
Acmella caulirhiza  116
Adenocarpus mannii  127
Aërangis gravenreuthii  155
Aframomum
    *chlamydanthum*  160
    flavum  159
    *hanburyi sensu Koechlin*  159
    zambesiacum  160
    sp. 1 of Fosimondi  160
    sp. 2 of Fosimondi  160
    sp. 3 of Fosimondi  160
*Afrardisia*
    *cymosa*  131
    *staudtii*  131
*Afrosersalisia cerasifera*  144
Afrostyrax lepidophyllus  123
Alangium chinense  108
Albizia adianthifolia  126
Alchornea
    floribunda  119
    laxiflora  119
Allanblackia gabonensis  75, 122, **Plate 4G**
*Allanblackia* sp. of F.W.T.A.  122
Allophylus
    bullatus  84, 143
    conraui  84, 143
    *sp. 1 of Kupe Bakossi checklist*  84, 143
*Alsophila*
    *camerooniana* var. *camerooniana*  162
    *manniana*  162
Amorphophallus sp. of Fosimondi  150
Amphiblemma mildbraedii  128
Anchomanes
    difformis  150
    *difformis* var. *pallidus*  150
    *welwitschii*  150
Ancistrorhynchus serratus  155
Aneilema
    dispermum  151
    silvaticum  87, 151
Angraecopsis ischnopus  155
Angraecum pungens  155
Angylocalyx oligophyllus  127
Anthocleista scandens  78, 128
Antidesma
    laciniatum var. laciniatum  119
    vogelianum  119
Antrocaryon klaineanum  109
Anubias barteri var. barteri  150

Ardisia
    kivuensis  131
    staudtii  131
Argocoffeopsis fosimondi  54, **57**, 80, 134, **Plate 7A**
Argostemma africanum  134
Asparagus warneckei  151
Asplenium
    paucijugum  162
    *variabile* var. *paucijugum*  162
Atractogyne bracteata  134

Baillonella toxisperma  143
Baissea gracillima  110
Basella alba  112
Batesanthus parviflorus  111
Begonia
    adpressa  70, 112
    ampla  113
    eminii  113
    *jussiaeicarpa*  113
    oxyanthera  71, 113
    oxyloba  113
    poculifera var. poculifera  113
    preussii  71, 113
    pseudoviola  72, 113, **Plate 4B**
    quadrialata subsp. quadrialata var. quadrialata  113
    schaeferi  72, 113
    scutifolia  114
    *sessilanthera*  113
Beilschmiedia sp. 1 of Bali Ngemba  125
Bersama
    abyssinica  130, **Plate 6F**
    *acutidens*  130
    *maxima*  130
Bertiera laxa  134
Bidens barteri  116
Biophytum talbotii  132
*Blumea crispata* var. *crispata*  117
Boehmeria
    macrophylla  148
    *platyphylla*  148
*Borreria princeae* var. *princeae*  141
Brachystephanus
    giganteus  67, 107
    jaundensis subsp. jaundensis  107
Bridelia
    atroviridis  119
    speciosa  119
Brillantaisia vogeliana  107
Browallia americana  145
Brucea antidysenterica  145

Bulbophyllum
  cochleatum
    var. cochleatum 155
    var. tenuicaule 155
  falcipetalum 156
  intertextum 156
  josephi var. josephi 156
  maximum 156
  nigericum 87, 156
  oreonastes 156
  *oxypterum* 156
  sandersonii
    subsp. sandersonii 156
    subsp. stenopetalum 156
  *tentaculigerum* 156
  *tenuicaule* 155
  *zenkerianum* 156

Calanthe
  *corymbosa* 157
  sylvatica 157
*Camptostylus ovalis* 121
Canarium schweinfurthii 114
*Canthium*
  *henriquezianum sensu Hepper* 140
  *hispidum* 136
  *kraussioides* 140
  *mannii* 137
  *setosum* 136
Carapa
  grandiflora 129
  procera 129
Carex echinochloë 152
Cassipourea malosana 134
Celosia
  *bonnivairii sensu Keay* 109
  isertii 108
  leptostachya 109
  pseudovirgata 109
*Celsia densifolia* 144
*Cephaelis*
  *mannii* 139
  *peduncularis*
    var. *hypsophila* 140
    var. *suaveolens* 140
Chassalia
  laikomensis 81, 135, **Plate 7B**
  simplex 135
  sp. 1 of Bechati 135
Chlamydocarya thomsoniana 123, **Plate 5A**
Chlorophytum
  comosum
    var. bipindense 150
    var. sparsiflorum 150
Chrysophyllum sp. of Fosimondi 144
Cincinnobotrys letouzeyi 79, 128
*Cissus*
  *adenopoda* 149

*mannii* 149
*rubrosetosa* 149
*Claoxylon hexandrum* 120
Clausena anisata **97**, 143
*Clerodendrum*
  *buchholzii* 124
  globuliflorum 124, **Plate 5D**
  silvanum var. buchholzii 124, **Plate 5E**
  *thonneri* 124
*Cnestis*
  *aurantiaca* 117
  *congolana* 117
  corniculata 117
  *grisea* 117
  *longiflora* 117
  *sp. A sensu Hepper* 118
Coffea montekupensis 81, **94**, 135, **Plate 7C**
Cola
  anomala 146
  ficifolia 146
  flaviflora 146
  lepidota 146
  verticillata 146
Coleotrype laurentii 151, **Plate 8D**
Colocasia esculenta 150
Commelina capitata 151
Connarus griffonianus 118
*Coreopsis barteri* 116
Costus letestui 152, **Plate 8E**
Crassocephalum
  bougheyanum 116
  montuosum 116
Crinum jagus 150
*Crossandra guineensis* 108
Crotalaria
  *acervata sensu Hepper* 127
  subcapitata subsp. oreadum 127
  sp. of Fosimondi 127
Croton macrostachyus 120
Culcasia
  *lancifolia* 151
  *saxatilis* 151
  scandens 151
  striolata 151
Cupressus lusitanica 161
Cuviera longiflora 135
Cyanotis barbata 152
Cyathea
  camerooniana var. camerooniana 162, **Plate 8I**
  manniana 162
Cylicodiscus gabunensis 127, **Plate 6C**
Cyperus
  fertilis 152
  tomaiophyllus 152
Cyphomandra betacea 145
Cyphostemma
  adenopodum 149
  mannii 149

rubrosetosum 149
Cyrtorchis
   arcuata var. variabilis 157
   brownii 157
   letouzeyi 88, 157
   ringens 157
   sp. of Fosimondi 157

Dacryodes edulis 114
Dasylepis racemosa 121
Deinbollia
   oreophila 85, 143
   *sp. 1 of Kupe & Bali Ngemba* 85, 143
   sp. of Fosimondi 143
Desmodium
   repandum 127
   uncinatum 127
Desplatsia subericarpa 147
Diaphananthe bueae 88, 157
Dichaetanthera africana 128
Dichapetalum
   heudelotii var. hispidum 119
   tomentosum 119
Dicranolepis vestita 147
Didymosalpinx abbeokutae 135
Digitaria pearsonii 153
Dinophora spenneroides 129
Dischistocalyx thunbergiiflorus 107
Discoclaoxylon hexandrum 120
Discopodium penninervium 145
Dissotis
   bamendae 79, 129
   *princeps* var. *princeps* 129
Dorstenia
   barteri
      var. barteri 130
      var. nov. of Bechati 130
Dracaena
   arborea 153
   camerooniana 153
   *deisteliana* 153
   fragrans 153
   phrynioides 153
Drypetes principum 120
Duparquetia orchidacea 126, **Plate 6A,B**

Elaeis guineensis 159
Elatostema welwitschii 148
Embelia schimperi 131
Emilia coccinea 117
Eragrostis camerunensis 153
Eucalyptus sp. 131
Eulophia horsfallii 158, **Plate 8F**

*Fagara rubescens* 143
Ficus
   ardisioides subsp. camptoneura 131
   *camptoneura* 131

sp. of Fosimondi 131
Floscopa africana subsp. petrophila 152

Galium simense 135
Garcinia
   kola 122
   *polyantha* 122
   smeathmannii 122
Geophila
   afzelii **95**, 136
   *hirsuta* 136
   obvallata subsp. obvallata **95**, 136, **Plate 7D**
Geranium
   arabicum subsp. arabicum **93**, 121
   *simense* 121
Gladiolus aequinoctialis 154
      var. *aequinoctialis* 154
Globimetula oreophila 128
*Gloriosa*
   *simplex* 151
   superba 151
Gnidia glauca 147
*Gongronema*
   *angolense* 111
   *latifolium* 111
Graptophyllum glandulosum 107
*Grewia*
   *barombiensis* 147
   *coriacea* 147
Guarea glomerulata 129

Habenaria weileriana 158, **Plate 8G**
*Haemanthus cinnabarinus* 150
*Hannoa ferruginea* 85, 145
Harungana madagascariensis 122
*Heckeldora*
   *latifolia* 129
   ledermannii 80, 129
   staudtii 129, **Plate 6E**
Heinsia
   crinita 136, **Plate 7E**
   *scandens* 136
Hekistocarpa minutiflora 136
Helichrysum cameroonense **92**, 117
Heterosamara cabrae 133
*Hirtella conrauana* 116
*Homalocheilos ramosissimus* 124
Hymenocoleus
   hirsutus 136
   *petitianus* 136
   subipecacuanha 136, **Plate 7F**
Hymenostegia afzelii 126
Hypolytrum heteromorphum 153
Hypselodelphys poggeana 154
Hyptis lanceolata 124

Impatiens
   burtoni 111

*deistelii* 112
filicornu 112, **Plate 3F**
kamerunensis subsp. kamerunensis 112, **Plate 3G**
letouzeyi 69, 112, **Plate 4A**
mackeyana subsp. zenkeri 112
mannii 112
sakeriana 70, 112
*zenkeri* 112
Indigofera
atriceps subsp. atriceps 127
*atriceps* subsp. *alboglandulosa* 127
Isodon ramosissimus 124
Ixora
*breviflora* 136
guineensis 136
*talbotii* 136

Jasminum preussii 132
Jollydora duparquetiana 118
Justicia
tenella 107, **Plate 3A**
sp. of Fosimondi 108

Kanahia laniflora 111
Keetia
hispida 'setosum' 136
mannii 137

*Lachnopylis mannii* 114
Lactuca
*capensis* 117
inermis 117
Laggera
*alata* var. *alata* 117
crispata 117
*pterodonta* 117
Lasianthus
batangensis 137
sp. 1 of Kupe checklist 137
*Lasiosiphon glaucus* 147
Leea guineensis 126
Leonardoxa africana subsp. letouzeyi 126
Lepisorus excavatus 163
Leptaspis
*cochleata* 153
zeylanica 153
Leptaulus daphnoides 123
Leptonychia
kamerunensis 86, 146
*sp. 1 of Bali Ngemba checklist* 86, 146
*sp. 1 of Kupe-Bakossi checklist* 86, 146
Leucas deflexa 124
*Lindackeria dentata* 121
Lindernia
nummulariifolia 144
senegalensis 144
Liparis
*guineensis* 158
nervosa var. nervosa 158

*rufina* 158
Lobelia
columnaris **91**, 114, **Plate 4C**
hartlaubii 115
*kamerunensis* 115
rubescens 115
Loxogramme
abyssinica 163
*lanceolata* 163

Macaranga occidentalis 120
Maesa lanceolata 131
Maesobotrya barteri var. sparsiflora 120
Magnistipula conrauana 74, 116
Mammea africana 122
Manniella gustavi 158
Marantochloa
*holostachya* 155
monophylla 154
Marattia fraxinea var. fraxinea 162
Mareyopsis longifolia 120
*Mariscus tomaiophyllus* 152
Marsdenia
angolensis 111
latifolia 111
Massularia acuminata 137
Medinilla mirabilis 129, **Plate 6D**
Medusandra mpomiana 128
Melinis minutiflora var. minutiflora 154
*Melothria*
*mannii* 118
*punctata* 118
Microcos
barombiensis 147, **Plate 8C**
coriacea 147
Microdesmis cf. puberula 132
Mikania
*capensis* 117
chenopodifolia 117
Mimulopsis solmsii 108
Momordica cissoides 118
Monanthotaxis sp. nov. of Bechati 109, **Plate 3D**
Musa spp. 155
Mussaenda
erythrophylla 137
tenuiflora 137, **Plate 7G**
sp. 1 of Fosimondi 137
*Myrianthemum mirabile* 129
Myrianthus
arboreus 115
fosi 59, **60**, 73, 115
preussii subsp. preussii 115
*sp. 1 of Kupe* 59, 115

Napoleonaea egertonii 78, 125, **Plate 5G**
*Neostachyanthus zenkeri* 123
Nephthytis
*constricta* 151

*gravenreuthii* 151
poissonii 151
*poissonii* var. *constricta* 151
Nuxia congesta **90**, 114

Olax
gambecola 132
latifolia 132
Oncoba
dentata 121
ovalis 121, **Plate 4F**
*Oreacanthus mannii* 67, 107
Oreosyce africana 118
Osmunda regalis var. regalis 162
Otomeria cameronica 137
Oxyanthus
formosus 137
gracilis 138

Palisota mannii 152
Panicum acrotrichum 154
Pararistolochia sp. of Bechati 111
Paspalum conjugatum 154
Pauridiantha
floribunda 138
paucinervis 138
Pavetta
bidentata var. bidentata 138
brachycalyx 81, 138
calothyrsa 138
gabonica 138
hookeriana var. hookeriana 82, 138
owariensis var. owariensis 139
Pavonia urens var. urens 128
Peddiea
africana 147
*fischeri* 147
*parviflora* 147
Penianthus longifolius 130, **Plate 6G**
Pentadesma grandifolia 122
Pentadiplandra brazzeana 133
Pentaloncha sp. nov. of Kupe Bakossi checklist **96**, 139
Pentas schimperiana subsp. occidentalis 139
Peperomia fernandopoiana 133
Perichasma laetificata var. laetificata 130, **Plate 6H**
Physalis peruviana 145
Phytolacca dodecandra 133
Pilea
*ceratomera* 148
rivularis 148
Piper
capense 133
guineense 133
Pittosporum
*mannii* 133
viridiflorum "mannii" 133
*viridiflorum* subsp. *dalzielii* 133
Platycerium stemaria 163, **Plate 1C**
Plectranthus

epilithicus 124
insignis 125, **Plate 5F**
kamerunensis 125
*luteus* 125
melleri 125
occidentalis 125
*peulhorum* 125
punctatus subsp. lanatus 77, 125
tenuicaulis 125
Pleiocarpa
rostrata 110
*talbotii* 110
*Pleopeltis excavata* 163
Poa annua 154
Podocarpus milanjianus 161
Poecilocalyx schumannii 139
Polycephalium lobatum 123
*Polygala cabrae* 133
Polyscias fulva 110
Polystachya
adansoniae var. adansoniae 158
bicalcarata 89, 158
bifida 158
*concreta* 159
cultriformis 158
fusiformis 159
polychaete 159
rhodoptera 159
tessellata 159
Prunus africana 134
Pseudagrostistachys africana subsp. africana 75, 120
Pseuderanthemum ludovicianum 108
Psorospermum
aurantiacum 76, 122
densipunctatum 122, **Plate 4H**
Psychotria
babatwoensis 82, 139
*calceata sensu Cheek* 139
camptopus 139
globosa var. globosa 139
lucens var. lucens 139
*malchairei sensu Hepper* 140
martinetugei 140
peduncularis
var. hypsophila 140
var. suaveolens 140
succulenta 140
sp. 9 of Kupe-Bakossi 140
*sp. A of Bali Ngemba checklist* 82, 139
sp. B of Bali Ngemba checklist 140
Psydrax kraussioides 140
Pteridium aquilinum subsp. aquilinum 162
Pycnocoma cornuta 120
*Pygeum africanum* 134
Pyrenacantha longirostrata 77, 123, **Plate 5b**
*sp. C sensu Keay* 123

Quassia sanguinea 85, 145

Raphidiocystis mannii  118
Rauvolfia mannii  110
Renealmia africana  160
Rhabdotosperma densifolia  144
Rhaptopetalum geophylax  144, **Plate 8B**
Rinorea
    dentata  148
    preussii  148
Rothmannia whitfieldii  141
Rubus
    apetalus  134
    *exsuccus*  134
    *rigidus* var. *camerunensis*  134
Rutidea
    decorticata  141
    olenotricha  141
    smithii subsp. smithii  141

Sabicea
    calycina  141
    venosa  141
    xanthotricha  83, 141
*Sakersia africana*  128
Salacia
    erecta var. erecta  115
    lebrunii  73, 115, **Plate 4D,E**
    lehmbachii var. pes-ranulae  74, 115
    pynaertii  116
    sp. aff. nitida of Fosimondi  116
Sanicula elata  148
Santiria trimera  114
*Sapium ellipticum*  120
Scadoxus cinnabarinus  150
Schefflera hierniana  68, 111, **Plate 3E**
Scorodophloeus zenkeri  126
Selaginella vogelii  162, **Plate 8H**
Senecio
    *clarenceanus*  117
    purpureus  117
Sherbournia zenkeri  141
Shirakiopsis elliptica  120
Solanum
    terminale  145
        subsp. *welwitschii*  145
        subsp. *inconstans*  145
        subsp. *sanaganum*  145
    welwitschii  145
*Solenostemon*
    *mannii*  125
    *repens*  124
Spermacoce princeae var. princeae  141
*Spilanthes filicaulis*  116
Sporobolus
    *indicus* var. *pyramidalis*  154
    pyramidalis  154
Stachyanthus zenkeri  123
Stanfieldiella
    imperforata var. imperforata  152

oligantha var. oligantha  152
*Stenandriopsis guineensis*  108
Stenandrium guineense  108, **Plate 3B**
*Stephania laetificata*  130
Sterculia tragacantha  146
Streptocarpus
    elongatus  121
    nobilis  121
Swertia mannii  121
Symphonia globulifera  123
Synsepalum cerasiferum  144
Syzygium staudtii  132

Tabernaemontana
    sp. of Bali Ngemba  110
    sp. of Fosimondi  110
Tarenna
    eketensis  142
    *flavo-fusca*  142
    fusco-flava  142
    lasiorachis  142
    vignei var. subglabra  142
Thalictrum rhynchocarpum subsp. rhynchocarpum  133
Thunbergia fasciculata  108, **Plate 3C**
Trema
    *guineensis*  147
    orientalis  147
Tricalysia
    discolor  142
    *mildbraedii*  142
    sp. B aff. ferorum  142
Trichoscypha lucens  109
Trichostachys petiolata  83, 142
    *sp. 1 of Kupe Bakossi*  83, 142
Tridactyle anthomaniaca  159
Tridesmostemon omphalocarpoides  144
Trifolium usambarense  127
Turraea vogelii  130

Urera gravenreuthii  148
Uvariodendron connivens  109

Vernonia
    hymenolepis  117
    *insignis*  117
    *leucocalyx*
        var. *acuta*  117
        var. *leucocalyx*  117
Veronica abyssinica  144
Virectaria procumbens  142
Voacanga sp. 1 of Bali Ngemba & Kupe  110

Xylopia africana  68, 109
Xymalos monospora  130

Zanthoxylum rubescens  143, **Plate 8A**
Zehneria scabra  118
Zenkerella citrina  126